The Cosmic Mystique

The Cosmic Mystique

Henry A. Garon

ORBIS BOOKS
Maryknoll, New York 10545

Founded in 1970, Orbis Books endeavors to publish works that enlighten the mind, nourish the spirit, and challenge the conscience. The publishing arm of the Maryknoll Fathers and Brothers, Orbis seeks to explore the global dimensions of the Christian faith and mission, to invite dialogue with diverse cultures and religious traditions, and to serve the cause of reconciliation and peace. The books published reflect the views of their authors and do not represent the official position of the Maryknoll Society. To learn more about Maryknoll and Orbis Books, please visit our website at www.maryknoll.org.

Library of Congress Cataloging-in-Publication Data

Garon, Henry A.
  The cosmic mystique / Henry A. Garon.
    p. cm.
  Includes index.
  ISBN-13: 978-1-57075-632-0 (pbk.)
  1. Cosmology. 2. Religion and science. 3. Catholic
Church—Doctrines. I. Title.
  BX1795.S35G37 2006
  261.5'5—dc22
                        2005028513

To my wife Marie, our children and extended families,
and the more than nine thousand persons who,
across a period of thirty-nine years, sat as students in my classrooms.
You, along with the ideas presented in these chapters,
represent much of what has been magnificent in my life.

Contents

Preface

After teaching physics for several years, I began keeping notes on my personal intuitions relating to what I understood as "the beyonds" of physics. These were individually dropped into a box as I went on teaching. After several more years I returned to the box, retrieved the notes, and began editing them under separate headings that eventually evolved into the present titles of my chapters. Further additions were made across a lengthy period of reading during which my personal intuitions and those of others were compared and recorded. Over the course of another twenty or so years, the manuscript was put aside, resurrected and revised many times. And now at last, seven years after my retirement, I offer this to my readers as an important personal closure of mine, wrapping up what I started a third of a century ago.

This book is written on an intermediate level for persons with a sense of wonder who love to gaze beyond the "mere stuff" of this world and ask enormous questions, who enjoy looking at creation in unusual ways. It is meant for whoever might acknowledge that baby crabs are precious, and then go on to recognize that the mama and daddy crabs, too, might be charming.

It is intended for those who might wonder about the significance of things and their interconnectedness throughout creation, who might delight in thinking of themselves as having been rescued from nothingness, who might ask if a dignity of a kind might be found in the phenomena of chance and accident. Also, it is meant for those with an imaginative spirit who might relate with the world to the extent of wondering at least occasionally what it would be like to fly like a bird, to be a volcano overlooking a lake, or to become an ear of corn for just five minutes.

Buried in matter, space, and time from deepest antiquity are endless secrets awaiting discovery. And beyond those hidden truths lies yet another realm, one that is even more mysterious: the world within ourselves where physical reality is assimilated and translated into meanings that are afterwards acted upon. It is a domain where we can come to recognize the commonplace as extraordinary in proportion to our willingness to listen and to flow with implications.

Nature demands that we physically interact with energized material throughout our lives. (Atoms have energies within their makeup; thus, everything everywhere is energized.) And spirit calls us toward reflecting in transcendent ways, bestowing meaning on things and making choices according to the values we see in them. There is clearly a mystique in things—*in all things everywhere*. Whoever is perceptive of meanings will come to sense this mystique in the odors of a forest in autumn, in the complex webs spun by spiders, in the cries of wild geese flying in formation, in the whistles of old-time steamboats, in the delightful voices of small children as they struggle to pronounce words correctly. Whoever loves the mystique of things need not search very far, because mystery is everywhere—in cloud formations, in sunsets, in the instincts of honeybees, and even in the random layout of pine needles fallen to the ground.

I write as one who has listened to nature and spoken of its behavior for a very long time. I am a scientist, a retired classroom physics professor who has lectured to more than nine thousand university students over a period of almost forty years. I now write about the world in an unconventional manner because the time has arrived for me to share my deepest thoughts with others. I am also a religious person, and as such I choose to write freely and openly, using my understanding of physics as a stepping-stone toward transcendental views of the world.

In these reflections I write of the world's haunting appeal, which draws us outside of ourselves as we repeatedly endow with meaning everything we come to know. I write at times about matter, energy, space, and time in ways that extend beyond the realm of physics, in ways that people almost never mention. For, indeed, there is something notably astonishing about the world behaving in the ways that it does: something wondrous and lovely about the awesome simplicity that underlies its complexities; something magnificent about the consistency in its ways, a consistency that stirs and nurtures our deepest sense of trust. After all, we do embrace the performance and behavior of this world, staking our lives on its dependability many times each day.

Starting from the baseline of the world's behavior as observed by scientists, I then move forward to consider ultimates, for I am convinced that life makes little sense without such considerations. I have avoided pantheism, the belief that God's essence is comprised of the substance of this world, or that the sum total of matter itself is the deity. However, I do insist that God's haunting presence is to be found in all of creation, and I have written at length about that mystery.

I hope my readers will not think of these reflections as proofs for the existence of God, or as scientific rationalizations of Christian beliefs.

Clearly, they are not. At no point do I mean to imply that scripture reveals the basic behavior of the physical world better than our textbooks of science. Nor do I want to imply that data of science supersede a good understanding of scripture. Science and religion are two faces of a single reality, and we ourselves are discerners of what the world presents to us. We are creatures who not only know, but who also hope. We repeatedly hope for this or for that from moment to moment throughout each day, and in fact most of these simple wishes are fulfilled.

Throughout these pages I suggest that matter and spirit are essentially compatible in the human psyche. I view the human mind itself as an outgrowth of that compatibility, and I assert that we can indeed sense transcendent mystery in things being what they are and behaving as they do. With this belief in mind, I will at times introduce my readers to the contemplations and visions of certain well-known writers, some of whom insist that the world is progressing toward something of unimaginable significance, of enormous proportions, so vast in depth and scope as to almost escape recognition. They maintain that reality as a whole is converging toward an ultimate culmination, proceeding as if in a crescendo of highest significance. One of these writers refers to this goal of convergence as the "Omega point," and in order to familiarize my readers with his thinking, I have devoted the final chapter to that topic.

The chapters are presented for the most part as a progression of thought. However, each is meant to stand on its own as a separate reflection. Therefore, readers need not hesitate to engage them in whatever order they wish. I only hope that my readers will enjoy reading them as much as I have delighted in writing them, a little at a time over a lengthy period.

Henry A. Garon

Acknowledgments

I wish to express my heartfelt thanks to my friends and colleagues who gave of their time and talents toward helping me during the lengthy period of composing the manuscript for this work. Foremost, I thank my wife, Marie Reynaud Garon, for her suggestions extending across decades.

I also thank the following who responded to my needs for information, advice, and encouragement. In naming them, however, I do not imply that they fully endorsed the views expressed herein: Lateitia Beard, Harold Baquet, Kurt Birdwhistell, Radu Bogdan, Clive B. Comeaux, Barbara Fleischer, Henry Folse, James Gaffney, Robert Gnuse, Joseph Grassi, William and Emilie Griffin, Albert Heine, Gary Herbert, Leonard Lassalle, Donald Martin, S.J., Donald Rodriguez, John Scurich, John Stacer, S.J., Manuel Correia, Edward Schott, S.J., Douglas Venne, M.M. and Muriel Cameron, RSCJ. My gratitude is extended also to the following, who by early 2006 were deceased: Gerald Clack, William Dych, S.J., Mary Helen Lorio, O. Carm., Ransom Marlow, S.J., Lurana Neely, S.B.S., Donald Walsh, and Youree Watson, S.J.

Special thanks go to my son, Boston College philosophy graduate Patrick Garon, who did the final proofreading and offered valuable suggestions. The drawings of Chesterton, Einstein, Teilhard de Chardin, Rahner, Kepler, and Newton are the work of Loyola University physics graduate Moctar Omar.

PLANET EARTH

This NASA satellite view of the planet Earth reveals many distinctive features that include the Amazon River, Hurricane Linda, the Andes, the Rocky Mountains, and the
edges of Antarctica.

CHAPTER 1

On Science and Aesthetics

"Ordinary reality is infinitely greater than our capacity to word it accurately."
—Wm. Bausch

From soap bubbles to galaxies, and even beyond, everything everywhere is alive with meaning. People everywhere are attuned to the basic realities of matter, energy, space, and time. No one can deny the truth of earth, sun, rain, and air being our benefactors. But not everyone discerns the finer, more elusive whisperings of nature. Such discernment requires a willingness to take the road less traveled, to slow down and listen attentively, to read with good judgment, to reflect in wonder on life and its meanings.

In a day and age when we address the visibles of this world with much fervor, all is not well between ourselves and the invisibles. This book addresses both the visibles and the invisibles, and focuses on where in our experience of "world"[1] the highest meanings are to be found.

Let us be willing to flow with implications, to proceed beyond the ordinary and visualize things from new perspectives, to view creation holistically. When doing this, we must cherish not only the external world of matter but also our inner worlds of mind and spirit that work together toward our discernment of what is of *quality* and what is of *worth*.

It has been this writer's privilege to associate with hundreds of scientists across the span of half a century, and never has he known men and women who are more fascinated than they. Most of them are deeply reflective. They seem rather slow in voicing their fascinations about things, preferring instead to simply observe, record, and privately delight in what the substance and situation of the world is telling them about itself.

Scientists view the world in terms of tiny discrete particles called atoms, and even tinier ones known as elementary (or sub-atomic) particles. They see the world, including light and even radio waves, as being "quantized."[2] They picture space as structured, as able to be distorted or "curved" by the presence of matter. When they announce their findings, they do so as factually and objectively as possible, often, it seems, with little emotion.

The information in this book is presented in ways that are different from the manner in which most scientists would write. Scientists of today generally regard aesthetic responses to creation as private and personal—as inappropriate for general discussion such as that at scientific meetings. It would be unusual for a scientist to speak about science as an artist might speak about art, in terms of beauty and graceful appeal.

However, disagreement with impassive ways of speaking about science has begun to surface, even within the scientific community itself. Certain scientists are insisting that the customary practice of thinking objectively about this world involves a certain illogic, for the human mind itself, as a vital part of creation, cannot be viewed as a separate and independent entity observing creation from the outside. Thus we, the dissenters, insist that, when gazing at the world, we are gazing also at ourselves with both our bodies and our minds included in the concept of "world." The implication of minds being so included leads to enormously significant considerations.

It is an exciting experience to navigate the world of the sciences, a realm of believability where value is rooted in what can be demonstrated. Scientists believe that nature is intelligible on levels deeper than the casual. They have devoted themselves to studying such things as laser beams, integrated circuits, rotating vectors, molecular motions, wave equations, x-ray radiations, and endless other things—and implicitly proclaim these to be deeply meaningful and worth their time. If anyone on earth has experienced what might validly be called "other-worldness" while addressing the ways of this world, it is the theoretical scientist. Not surprisingly, theoretical scientists have been described by some as "having their heads in the clouds." I myself have great respect for theoretical scientists.

Scientists are persons enraptured by the world's substances and the interrelations found between them. They spend their professional lifetimes delving into realms of order and disorder, atomic valences, relativity, fluid flow, galactic motions, molecular structures, and such. They dream up mathematical expressions to describe what they have found. Their equations are shorthand representations of relationships, ways of "extrapolating" and saying very much with just a few symbols.

However, one need not be a scientist to deeply appreciate creation, for everyone encounters creation in highly personal ways at every turn. Each of us strives in our own personal way to discern the implications of whatever segments of the world we engage. With the passing of time, we increasingly reflect on the meanings of things everywhere. As we get older, we ponder such things as planets, mountains, skyscrapers, ocean waves, mountain goats, and also the so-called "lesser" things such as rocks, weeds, grains of sand, and dried-up branches decaying near ravines and gulches. From childhood we have been entranced by things, many of which have woven their ways into our inner worlds of expectation and desire. In later years we must learn to sooner or later let go of them, even after having grown to realize that all things everywhere are astonishingly extraordinary.

Why do we consistently judge certain things to be worthy of our close attention and give them our time? Why do certain people go to the trouble of replacing the blades on their windshield wipers when they could easily afford to purchase new wipers? Is it only in order to save money? Or might there be other reasons? What is the source of the lure of yo-yos, gyroscopes, calculators, and electric light switches on our walls? What do they offer us? What is the promise that they hold out to us as we reflect on them and discover in them patterns that hold true regardless of time or place?

Throughout this book we shall undertake a pursuit of meaning, maneuvering at times like mavericks in and out of unexpected places. We shall navigate not only the externals of things but also their inroads and "inscapes," including the arenas of the human spirit that respond to the charisma of things, people, and places. We shall weave a tapestry of a sort, engaging at times in poetic science and even in a poetic theology, a kind of thinking recognized by the great theologian Karl Rahner (1904–1984), who cited the need for it in today's world.

Viewing things from both scientific and religious perspectives, we will repeatedly consider ways in which the worlds of science and religion blend. And, further, from the Christian point of view we will consider how our relationship with God is intrinsically affiliated with our proper intimacy with "world," inasmuch as God has chosen to *personally identify with this world* through Christ.

It seems like a great oversight when a person who philosophizes about nature is never heard speaking about it in the framework of spirit. An image comes to mind at the moment, a picture of a highly acclaimed piano tuner who, while striking and tuning individual notes one at a time throughout his life, never listens to an entire musical composition. So it is when we neglect to mention the transcendent dimensions of nature while studying nature's ways. We thereby risk ignoring so very much of the colossal cosmic orchestration of which, let us not forget, we ourselves are a transcendent part.

CHAPTER 2

Fantastic Good Fortune

"The real voyage of discovery consists not in seeking new landscapes,
but in having new eyes."
—Marcel Proust

Years ago, while I was working in the oil fields near the mouth of the Mississippi River, my co-workers and I often sent measuring instruments and earth-sampling tools deep into the newly drilled holes. We obtained actual samples of the earth taken from various depths, some from as deep as three miles below the surface. The specimens we obtained included beautiful white salt, hard-packed sands saturated with oil, and tiny fossil sea shells that told us much about the biological and geological processes that must have occurred there eons ago.

The work we were doing sometimes evoked in us unusual feelings of kinship with the inner substances of the earth. For example, we often took the earth's temperature at various depths—say three miles down, where we found it to be around 260 degrees Fahrenheit! Rather hot! These oil well instrumentation experiences made me feel intimately connected to this huge globe and its hot interiors.

There were other times, also, when I could feel such connections. On one occasion, I had the unusual experience of sensing the roundness of the earth by way of radio waves. This happened one night when radio propagation conditions were exceptionally good. I was in New Orleans, using an amateur radio station to chat with an Australian friend, when both he and I noticed something peculiar about each other's voices: each word we spoke seemed to carry an echo of its own. The cause of those glorious echo-like sounds soon dawned on us. Under that night's exceptional conditions, our waves were traveling *in both directions* around the planet. And because Australia is not exactly on the opposite side of the earth from New Orleans, my words were reaching him at slightly different times. They came to him by the shorter path across the Pacific and

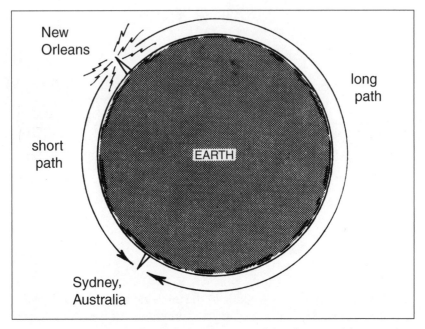

"Our waves were traveling *in both directions* around the planet. And because Australia is not exactly on the opposite side of the earth from New Orleans, my words were reaching him at slightly different times."

also by way of the longer and opposite path across the North Atlantic, Russia, and southern India. The longer path accounted for the slight delay in arrival, which is why we each heard the other's syllables twice.

The sentiments evoked by this experience are hard to describe. In a manner of speaking, it was as if *our voices were ringing in the roundness of the earth,* as if we were extending our arms in both directions around the planet, thus personally sensing the size and shape of the huge sphere ...all by means of our two sets of radio waves. What an unforgettable and wondrous experience that was!

My colleagues and I at Loyola University have another way of experiencing a certain reality of our planet. Vibrations from distant quakes, on passing through our area, cause our delicate seismographic instruments to swing back and forth. The very slow swinging of our seismic pendulums—about one swing every twelve seconds—tells us that huge masses of earth materials are undergoing vibrations. Along this avenue of thought we have acquired further insights into the enormous mass of the planet as a whole. The process is analogous to that of judging the size of a boat by how rapidly it rocks from side to side when traveling through

rough waters. Small boats rock rapidly, whereas huge ships rock quite slowly.

It is as if our seismic instruments have become extensions of ourselves. So huge is the mass of the earth as a whole that if it were to be suddenly struck by a large asteroid, it would take approximately fifty-three minutes for the earth to make one complete wobble. And, if struck in that way, it would continue to wobble very slowly for a long time—somewhat in the manner of a bell that continues to ring for a time when it has been struck just once.

Yet another unusual understanding of the earth came my way recently, when a Canadian physicist related to me his early experiences as a coal miner in Ontario. The work took him to depths in excess of seven thousand feet, to places where fresh air had to be piped in so as to counteract the intense heat.

"I would guess that it must be pretty silent that deep in the earth," I said to him.

"No, indeed!" he was quick to respond. "You'd be surprised at the noises to be heard down there!"

He went on to describe how miners working at such depths are able to hear "earth noises," the eerie sounds of squeezing and cracking as layers of earth slip and slide over one another. The cause of this is as follows: The shape of the earth is somewhat distorted by the gravitational pull of the moon. Instead of being a perfect sphere as we often visualize it to be, the earth bulges very slightly, somewhat like an egg. As the moon continually moves around the earth, this bulge continually readjusts, pointing always toward the moon that causes it. This results in layers of materials within the crust of the earth continually slipping and sliding, one over the other.

It is normal for us as imaginative persons to experience deep feelings when we reflect on the various materials and processes that we find in our world. Feelings such as curiosity, gratitude, amazement and delight arise within us as we contemplate the blueness of the sky, the beauty of raindrops, the wonder of photosynthesis, the delicacy and strength of webs spun by spiders, the sparkling colors from sunlight reflected in dew, and endless other marvels. Repeatedly, we soar into new dimensions of consciousness while addressing the same old world with new awareness.

One of the rewards of studying the physical sciences is that they present us with a wealth of rich images and visualizations. Elegant and beautiful concepts such as energy-level transitions, the time rate of change, the gradient and curl vectors, magnetic resonance, the warping of space-time, and invariance in the speed of light are commonplace in science. Physical scientists are highly imaginative persons who routinely visualize the world in unusual ways.

When I was teaching, I was often fascinated by the reactions of my physics students when, in order to startle them a little, I asked, "Do you realize that, as inhabitants of the earth, you were born in a gravitational trap? Have you ever thought of yourself as a prisoner of the earth?" They seemed amazed at this . . . and at the idea that our sphere orbits a nuclear furnace, the star that we call our sun—which we never see as it is at the moment we observe it but, rather, *as it was about eight minutes ago*. For it requires that amount of time for the light from the sun to reach us on earth. Somehow, it was through statements such as these that the strangeness of our cosmic situation began to dawn on them.

Strange-sounding questions can be asked about anything and everything, from popcorn and peanuts to black holes in interstellar space. In what ways would our solar system be different if gravitation behaved in ways other than the way it is known to behave? What would our world be like if water froze at 20 degrees Celsius instead of at zero degrees? What would life be like if the atmosphere contained oxygen in the ratio of one part in ten instead of one part in five? How would the structure of animals have evolved if the pull of gravity at the earth's surface had been (throughout the ages) twice as strong as it presently is? What would trees, honeycombs, and birds look like today had this been so?

Imaginative questions such as these are essential and ought to be entertained, for the habit of contemplating things as they might have been leads to awareness of the wonder of the world as it actually is. This habit can in time lead us deeper into what classical philosophers have called an "ongoing awareness of being" as we wonder about that which is knowable.

Reflective persons, in time, become aware of a quality of strangeness about everything. Through fresh realizations on our part, the world itself is born anew within our consciousness. No longer are things, situations, and persons taken for granted, for we ourselves become new creations when undergoing radical new experiences, such as when we experience a narrow escape of some sort. An example of a very narrow escape occurred in my family during World War II.

It happened that my eldest brother Allen, then a U.S. Navy officer, was flying in a commercial airliner from New Orleans to San Francisco. During a stopover in El Paso, he was "bumped" from the fully loaded plane so that others holding higher priorities than he could board. He, of course, was keenly disappointed at the ensuing delay. However, the following day he learned that the plane from which he was bumped had crashed on its approach to San Francisco. The pilot had made an erroneous turn in heavy fog and the plane had slammed into a mountain, killing everybody on board.

The incident caused a dramatic shift in the way that he would thenceforth view his life. For him, each ensuing day would become a dramatic prolongation of life, a time given to him to observe what he would have missed had he remained on that particular plane.

As to the countless physical phenomena deserving our appreciation, let us now consider gravitation. The planets, stars, galaxies, and other celestial objects are subject to this mysterious and universal attractive force by which all things throughout the universe pull on one another. Gravitational forces account for both individual planets and for planetary systems. For instance, on the solar scale of vast masses and distances, gravitation's enormously strong forces hold the sun together, hold the earth together, and keep the earth in its orbit around the sun. In truth, our whole world of living things relies on gravitation consistently holding the earth at a life-enhancing distance from the sun.

Yet, locally speaking, the forces of gravitation pulling on things—automobiles, people, garbage cans, dogs, bowling pins—are weak enough to allow us to navigate without being drawn together with nearby objects in massive pileups. Because our individual masses are small, gravitational attraction, although present, is not noticeable between ourselves and objects other than the earth. Even the most massive of local objects, such as large buildings and mountains, lack sufficient mass to capture us or noticeably restrict our motions. Thus, freedom of navigation along the surface of the earth is not only possible, but also guaranteed.

Gravitation is at work within the sun, holding the gases together in a sort of balancing act between outward explosion and inward collapse. Gravitation draws the hydrogen atoms together and squeezes them in an interaction known as "fusion," which results in their emitting many kinds of radiation, including heat. The process, roughly speaking, may be compared to squeezing a water-soaked sponge. As one squeezes the sponge into itself, the water inside it comes squirting out. Similarly, gravitation at work within the sun squeezes the hydrogen atoms together, instigating nuclear reactions that spew out radiation in all directions.

Not all of the emissions from the sun are beneficial to humans. Most of the harmful ultraviolet radiation is absorbed by the earth's atmosphere. Yet there are other types of radiation, consisting of deadly, high-energy, electrically charged particles that, in time, would destroy all life on earth. Fortunately, these particles are deflected away from us by the earth's magnetic field, which functions as a protective umbrella of a sort. Thus, the earth's magnetism shields us from death-dealing particles emanating from the very thing that nurtures our life—the sun!

But from whence does the earth's magnetic field come? There is now a growing realization that it is probably generated by metallic fluids—

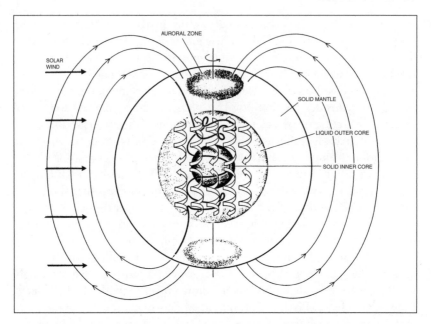

This diagram above illustrates how the solar wind, when approaching the earth from the left, encounters the earth's magnetic field. The field then acts as a protective umbrella, deflecting lethal charged particles away from the earth. Also shown is a depiction of the (imagined) way that the magnetic field is generated from molten metals circulating within the earth. Thus, the protection we enjoy originates from deep within our own planet. (Art and caption © George V. Kelvin)

molten iron, in particular—moving deep within the planet in a flow that is caused by the rotation of the planet. Our protection from the deadly emissions of the sun, it seems, originates from deep within the planet itself. Fantastic good fortune, we must admit!

Some questions worth asking are: What would life on earth be like if the earth did not rotate? What about day and night? Seasons and crops? Airline routes? What would the environment and life itself be like on the continuously dark side of the earth?

The role of gravity enters the picture once again in regard to *escape velocity*. This term refers to the speed at which an object moving directly away (upward) from a planet would break free of its gravitational bonding to that planet. Were there no atmosphere to slow things down, the velocity of escape from the earth would be about seven miles per second or twenty-five thousand miles per hour.[1] This tells us how fast we would have to throw an object upward (with superhuman arms, of course) so that it would never fall back to the ground.

Were the earth's gravity much less, or its temperature much higher, air molecules undergoing their ordinary vibrational motions would exceed the escape velocity for the earth and thus leave the planet. The earth would lose its atmosphere and become a lifeless sphere, similar to the moon. The moon has no atmosphere because its gravity is insufficient to retain gas molecules moving in thermal vibration from the sun's heat. Such molecules can easily reach speeds greater than the speed of escape from the moon.

These examples are but a few of many life-or-death considerations at issue in the circumstances and situation of our planet and its star, the sun. If we think of each of them as an either/or event, then our marvelous good fortune in being alive and thriving on planet Earth readily becomes apparent. In a manner of speaking, we have repeatedly beaten the odds, like someone having flipped a life-versus-death coin numerous times, and every single time having had the coin land on the side of life. There are also many less extreme "might-have-beens" which merit consideration:

As a weak, selective scatterer of blue light, air itself accounts for our sky's appearance. It is the blue light scattered by mile after mile of air that gives our atmosphere the soft, mild hue we see in the "blue sky." Let us, however, imagine the consequences for life had the physical properties of air molecules been different, so as to make them strong (rather than weak) scatterers of light. What effect would this have had?

One example of an intense scatterer of white light is fog. Fog limits our ability to see at a distance. But what if the *air itself* were a strong scatterer of white light? What, indeed, would life be like if air constantly appeared as foggy, even in the best of weather? How might this have affected human life in the course of history?

Let us suppose, for example, that normal visibility through the cleanest air was no greater than one hundred feet. What effect would this have had on our knowledge of astronomy? On highway transportation? On sailing? On photography? On the invention of the airplane? On sports such as football and baseball? What psychological effects would limited visibility have had on humans who would have been unable to view mountains and lakes in the way that we presently do? How would we survey land?

What might it mean to have an earth that is lifeless and devoid of creatures such as humans who can address creation in search of meanings, who can marvel over what they come upon? If we turn our telescopes skyward, we do in fact discover such planets, where atmospheres are hostile to life as we know it. These are still excellent creations, but excellent in ways other than suitability for human life.

It is very human to be impressed by rarities and scarcities. And it is also significant that we recognize our good fortune by way of abundances and prosperities. The fact that the whole cosmos is there before our eyes with ourselves as part of it is notably wondrous to anyone who is receptive to meaning.

Indeed, we are privileged to be aboard planet Earth, to draw upon its nutrients, its atmosphere, its energies, its seasons. We are fortunate to recognize it as offering us life-enhancing possibilities. To date, we know of no other such planet in the universe. Furthermore, the possibility of other planets on which we might someday make a landing and thrive seems so extremely remote as to hardly merit serious thought, at least for now.

It is customary for persons of religious faith to visualize themselves as having been spiritually saved, but few of us extend that conviction to include our planet. When we consider the many narrow escapes on planet Earth in favor of life, a new realization emerges. It becomes fully proper for us to visualize the earth itself as an entity that has been saved many times over. In a very real sense it passes that natural salvation on to us, its inhabitants. Repeatedly and in differing ways, even before the dawn of humans on its surface, the earth was rescued, exempted, and preserved from conditions that would have ruled out life. We ourselves are the highest outcome of that natural salvation. It's time we celebrate!

Note

1. The concept of "escape velocity" is always visualized and defined under idealized conditions where no atmosphere exists that would hinder the motion of a projectile. In the presence of an atmosphere, a greater velocity would be needed to cause an object to permanently escape its planet.

For Discussion

1. Name several things in today's world that you view as "extensions" of yourself.

2. Have you ever had a particular narrow escape that caused a shift in how you view your life?

3. In what sense can it be truthfully said that your entire life has been a narrow escape?

CHAPTER 3

Endless Extravaganza

People in olden days were profoundly watchful of the nighttime skies, attaching much significance to the stars. So strong was their search for meaning in the random distribution of the stars that they came forth with something new: they visualized certain of the more prominent stars to be connected by straight lines. Out of this came their constellations: groups of stars which to them represented such things as an eagle, a warrior, a lion, a queen on a throne, a dipper, or a southern cross. Much of their good work remains in use today. Whoever utilizes the Big Dipper in locating the North Star is well aware of that.

These are exciting times to be in astronomy. Highly unusual discoveries are being made. Several years ago newspapers featured stories of our discovery of a string of galaxies seven hundred million light-years long, extending halfway across the sky at an average distance of two hundred million light-years from the earth. It is thought to be the largest structure ever observed.

Not too long ago, a photograph of the galactic center of our Milky Way was obtained. It featured spiral arms extending outward from a central point, and speculation runs high that we are seeing matter falling back into a mysterious black hole at the center of the galaxy. That photograph was certainly among the most amazing ever to appear in print.

In 1987 a *supernova*, or exploding star, was discovered in the southern skies, the first to be seen by the naked eye in centuries. Astronomers are now measuring the kinds of radiation it emitted in order to better understand the mechanics of star explosions. In this enterprise, the excitement of these astronomers is focused on an event that, while appearing to have occurred in 1987, in fact took place about 170,000 years ago.

Looking outward from our solar system, we observe a fantastic number of stars—nuclear furnaces burning for eons. Our sun itself is a star, behaving somewhat like a wobbling ball of Jello as it vibrates in response to explosions occurring within it. Astronomers estimate that there are about 100 billion stars in each of the 100 billion visible galaxies. That multiplies out to approximately 10,000 billion billion stars in the observable universe. In scientific notation, that would be written as 10^{22} stars. So distant are some of the observed galaxies that, in spite of light traveling at 186,000 miles per second, we are presently seeing them as they actually were before our earth was formed. Clearly, our imaginations can hardly fathom the enormity of the space-time dimensions that stretch out before us. What deeply meaningful things are at work every time we look at the stars!

Astrologers of antiquity often gazed at what they thought was a single bright star. Their successors, the astronomers of later centuries, gazed at those very same "stars" through high-resolution telescopes and found that they were actually clusters of ten thousand or more individual stars. These globular clusters were so far away as to appear to the unaided eye as a single bright star. Discoveries such as these challenge our very imaginations. In our cities we miss most of these marvels. If we want to get a clear view of this stellar extravaganza of which we are a part, we must travel out into the countryside, fifty or more miles away from city lights. There, on a clear moonless night, we can see awesome sights: countless millions of stars displayed across the apparently endless reaches of outer space.

Have you ever imagined yourself as a passenger aboard a space ship traveling beyond the outermost reaches of the stars? In a situation such as that, everything would be located behind you. Suppose, then, that someone were to inquire about your destination—where you were going. What might be your answer? Every time you have traveled in this life, it has always been with awareness of yourself moving away from certain things and toward other things. But now, located beyond the farthest of stars, toward what would you be proceeding? On this question there is presently a great deal of speculation in the branch of physics known as cosmology, and much of the speculation has to do with space curving back on itself. An understanding of the apparently endless leads us to the very edge of what can be imagined.

For now, however, let us move in the opposite direction, namely toward that of smallness in the cosmos, a topic that we will approach from the viewpoint of aesthetics.

When we drive along highways, we may be saddened by the sight of junk, such as rusty, abandoned automobiles, empty beer cans, and other discarded objects lining the road. Their presence there indicates that some of our fellow citizens are irresponsible and certainly have little concern for the aesthetic experience that others might derive from things of beauty. Indeed, it takes some imagination on our part to come to terms with these disappointing experiences. But suppose we remind ourselves that, if we were to reclaim such ugly objects, taking them into a physics

Often what appears to be a single bright star, on closer observation through high-resolution telescopes, turns out to be a globular cluster of individual stars. Pictured in the NASA photograph above is Omega Centauri, the largest globular cluster in our Milky Way galaxy. It contains about ten million stars. Stars in globular clusters are generally older, redder, and less massive than our sun.

laboratory and shooting narrow beams of x-rays through them, we would observe astounding orderly structures within them. If we then allowed the emerging x-ray stream to fall upon photographic film, we would obtain x-ray diffraction photographs displaying surprising patterns of symmetry and beauty within those rusty and bent metals. The patterns would reveal the presence of atoms in geometrically perfect spacings and alignments within the crystals that compose the metals. Strangely, our experiment would suggest that we normally fail to see beauty in discarded things because we do not look at them closely enough.

We know much today about the world of the very small. We are aware that atoms exist in over a hundred varieties. Utilizing special microscopes, we have recently progressed to the point of directly observing clusters of individual atoms, tiny as they are. Atoms are so small that across the thickness of a copper penny, for example, there are about three million layers of copper atoms. (Three million is approximately the number of people required to fill thirty large football stadiums, each with a capacity of 100,000).

Chemists long ago discovered that recurring similarities exist as we proceed from simple atoms toward atoms of greater complexity. The well-known periodic table of the chemical elements is an outcome of their work. Physicists have probed even deeper into the realm of atoms

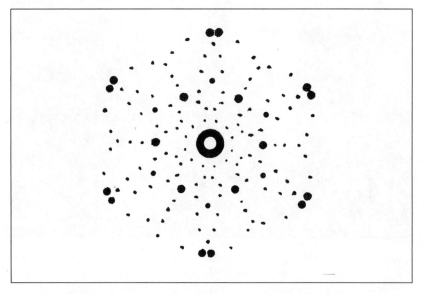

Depiction of a typical pattern produced on a photographic plate by an x-ray beam that has passed through a thin sheet of metal. (Author's sketch)

by smashing them apart and using Geiger counters, cloud and bubble chambers, photographic film, and other such detectors to examine the flying debris. Some express their doubts as to whether the recognized elementary particles on our ever-increasing list (at present more than two hundred) are truly elementary. Among the questions that persist are these: Is there an end or a bottom to smallness of size (or mass) in the cosmos? Are there entities in this universe that will, in principle, remain forever indivisible?

When we reflect in this manner, we find ourselves situated between two poles of apparently endless extension, both *outward* and *inward*. Our very ability to marvel is part of the issue here, for apparent endlessness is a consideration of utmost importance as we explore the philosophical question of who we are.

A tentative answer to our question of self-identity seems to be that we are intelligent, reasoning creatures who exist, physically speaking, between boundless largeness and unlimited smallness. But beyond the physics of that consideration lies a further answer. We are creatures who can reach out beyond arm's length even as we stay at home. We can imaginatively extend ourselves beyond our physical limitations of space and time as we unfold, prolong, and project ourselves by means of our mental fantasies and imaginative journeys. We can allow and dispose ourselves to be carried away in transitions of spirit. We can move into other worlds, particularly into the realms of meaning, hope, and love that draw us beyond what is immediate, observable, and measurable.

Even as we specialize in one or more fields of science, we remain persons who are called to integrate, harmonize, and reconcile our specialties with the vastness of an all-inclusive creation that draws us into metaphysical dimensions. We are called to love, to serve, and to surpass the purely material. We are invited in spirit to move beyond our limitations, to fly out, to overshadow, and to transcend the physical world as we commonly know it.

In consciously observing things individually and collectively in unlimited arrangements and disarrangements near and far, we open ourselves to their allurements. We respond with personal concern toward things, even to some extent toward what we choose to ignore. Things by their nature draw us beyond our immediate selves. Yet, paradoxically, the unparalleled experience of soaring beyond the merely material, of "outdistancing" the world by way of our spirit, is an outcome of our thriving in a situation of cosmic poverty. For ours is a world where we humans, when psychologically reaching for our stars, do so along the baseline of our limitations. Among these limitations are restrictions in distance, quantity, quality, energy, and diminishing health over time—the many givens of our awfully finite world.

Yet, in our cosmic condition of life, the finite and the endless are intertwined. Besides thinking of specific things as massive, or dense, or brittle, or sizable, we speak of them also in ethereal dimensions—as "coming in handy," as being "life-promoting," or wonderful, or truly "worth keeping." And we include ourselves in this transcendent language by proclaiming ourselves to be everlastingly grateful for the countless gifts of creation. So it is that, even in addressing finite things, we move beyond them, repeatedly losing ourselves in their charisma.

Indeed, we are creatures who strongly participate in the material. Yet, we transcend this world of the physical, navigating also the boundless worlds of the human spirit. Thus are we are led into apparently endless engagements in the beyonds of this world.

For Discussion

1. Someone once exclaimed that "everything everywhere is gift." Do you agree?

2. If, as scripture says, creation is good, what then does it mean when we say that certain things are wonderful whereas other things are horrible?

3. Were it possible for you to safely travel throughout the cosmos at super-high speeds, where would you most like to go? What would you most like to avoid?

CHAPTER 4

Toward Infinity
in Our Perceptions of the World

"Infinity is the sum of the cross ties
multiplied by the voltage in the third rail."
—Chicago Motorman's Manual

Ontologically speaking, we have strong reasons for believing that the world exists in certain knowable ways, but with countless creatures sensing it in ways that are unique to themselves. Never in our experience of knowing the world can we put aside our humanity and speak of the world as if we were wholly objective observers, for the world is colored immensely by ourselves—by our senses, our hopes, our goals, and our past experiences.

I stumbled across this realization one day in a rather strange way. While watching ducks in a pond and identifying with them just a little, I suddenly realized that they do not experience water in the same way that humans do. They do not feel the water against their skin as we do, for their oily feathers isolate their skin from the water. Their legs and feet are covered with a substance like that of our fingernails, which have no feeling. Neither can their bills feel water as our lips do. Thus, if ducks could describe water based on their experience of it, they would certainly relate their impressions in ways far different from ours.

When philosophizing about nature we must keep in mind that our science books, as expressions of a human interpretation of the world, are tailored to human readers. Like chairs, drinking cups, and bicycles that have been engineered to fit the bodies of average-size humans, our physical sciences are also expressed according to what humans can assimilate and understand. All writings are slanted and relative, at least in that sense.

We might do well to imagine, for example, how a horsefly with compound eyes would describe from the horsefly-point-of-view the curved

19

paths of baseballs in flight, or how a creature lacking eyes would write about the physics of rainbows. Indeed, science literature written by non-human creatures (sometimes referred to as "extraterrestrial science"), if that were possible, would have little meaning for us unless it were translated into terms understandable to humans. This is because we are bound by the parameters of our humanity when observing and describing what we experience—all of which, when taken together in this life, constitutes what we understand as "world."

We should also bear in mind that our ideas about the world are never identical with the realities themselves. After all, what is known by observation is always distorted to some extent by us who observe. That being said, one of the characteristics of science is that its observations about the world are usually made in ways that are as free as possible from unnecessary subjectivity.

Our ability to know is, indeed, a great mystery. Aristotle taught that to know a thing is "to become it" in a non-material way. He pictured the act of knowing as a relational activity, an "assimilation of otherness" by the observer. Let us keep Aristotle's idea in mind when studying, as we do, the complexities of the cosmos, for in knowing physical realities we enter into very personal relationships with them. And when we relate intimately with things and people, assimilating something of their most fundamental being, that being of theirs remains with us indelibly. On this point, Hassett, Mitchell, and Monan wrote:

> Being (or existence) is the lord of the intellect; and whenever we make a judgment that we know is true and certain, it is because being is so presented to the intellect that reality determines, forces and constrains us to assent to it. There can be no turning back at this point because, in the field of knowledge-content, the intellect is the servant and reality is the master.[1]

It is helpful to understand that our everyday awareness of things represents a kind of natural transcendence, a soaring beyond the immediate. When we observe someone, for example, we do so in response to images of that person being formed by light impinging on our retinas. Within our brain we assimilate the electrochemical signals from our retinal images and, psychologically speaking, *move beyond* the images themselves in coming to recognize the person. Thus, we never say to someone, "I sense inverted images of you on my retinas, and they indicate that you are really out there and right-side-up!" Instead, we assimilate and transform the images immediately into knowledge of the person viewed. We take the natural leap and simply say, "I see *you*." So, too, with our other senses. We never tell a singer, "I know by the frequency with which my ear drums are vibrating that you are humming at a high pitch." Instead, we immediately advance beyond that basic perception and say, "I hear *you* humming at a high-pitch." At this basic biological level, transcendence is something we share with the animals. They, like us, skip the intermediate steps in the awareness-gathering process and respond to their perceptions of things with high immediacy, often with an acuity greater than ours.

But, as humans, we enjoy vastly higher modes of transcendence than do the animals. Our knowledge of things stimulates our imagination, which somehow leads us endlessly into new dimensions. Once we know a thing, then in principle we also know by way of imagination and extrapolations an endless number of similar things, as well as never-ending variations of the original thing.

If we show people a flower, they can, if they so decide, immediately visualize a million flowers identical to the one we showed them—or a million flowers that are different in some way from one another and from the one we showed them! As humans, we tend to relate our new knowledge of a thing to our previous knowledge of things. We relate our newly acquired knowledge, also, to images of possible things (some of which may not actually exist).

Our individual acts of acquiring knowledge serve in endless ways to bring together people, things, and situations, past and present, within our minds. We assimilate and participate in the charisma of things, working them together in relational ways as if to create a network of a sort. We do not know things in isolated ways. For example, in recognizing a cow, we do so with the awareness that a cow cannot be fitted through the eye of a sewing needle—and that a sewing needle, in turn, cannot be used for pole vaulting, which, in turn, cannot take place in outer space, and so on.

Thus, when knowing people, things, and situations, we take all sorts of imaginative and relational journeys into the wonders of their individual and collective portrayals of being. We evaluate them, elevate them, and dismiss them. Often we succumb to their invitations, being carried away by their charisma and entering another world that can be positive or negative.

The result is a transcendent newness (as distinguished from physical newness) repeatedly burgeoning within the cosmos through humans. That is, our ongoing harvesting of knowledge entails ongoing changes within the cosmos as a whole—a revolutionary concept, to be sure, yet one that happens simply and repeatedly as follows.

We continually recognize all kinds of things and people, defining and redefining them within our personal worlds. In this way they become, at least from our point of view, reshaped creations with fresh significance. Subconsciously, we assign them value-meanings according to how they promise us something, how they "come in handy," the extent to which they seem useful or useless to us, and so forth. Their appeal to us and our need for them are important ingredients in this adventurous engagement of ours.

Our very acts of knowing, visualizing, and responding to things and people are basically creative in essence, extending as they do throughout the whole of nature. After all, to know something is also to dignify it by assimilating it into our psyche, and this can be correctly understood as inducing a metaphysical change in the cosmos as a whole. Whenever a change takes place within us, the cosmos itself undergoes an important transition, inasmuch as we are a uniquely significant part of the cosmos.

Furthermore, from a theological standpoint, our attention to anything in the cosmos represents a kind of coming of God into our midst, for one cannot pay attention to creatures apart from expectation. All observations, including scientific investigations, are hope-expectation engagements wherein we continuously assimilate something of Being by way of what we observe.

Thus, to simply turn our attention toward things and people is to call them forth to fuller life within our awareness. It is a continuance of the naming of the animals alluded to in the Book of Genesis (2:20), a kind of blessing we bestow on creatures in recognition of the God-likeness—their quality of existence, for example—that we see in them. Creatures everywhere go about their business of revealing unique modes of God-expression, triggering within us various experiences corresponding to their charismas—their loveliness, their flavors, their tones, their vigor, their inscapes, their particularities—in the settings of what we call "world."

In science we understand things as having been touched by countless other things that have brought them to where they presently happen to be. Many scientists declare that something of the history of the cosmos is present in each of its atoms, and that each of the things we study tells us something about where it has been and where it is going. Thus, to be engrossed with any one creature is to be in touch also with the whole of creation. Traces of the presence of everything everywhere reside in every particular thing.

We can focus our attention on a particular water molecule and reflect on how many times in the past this particle has fallen as rain, only to rise by evaporation and then return again to the earth. How many round trips has it made between earth and sky? Between America and Europe? Between Asia and Africa? What was its location at 3:09 AM on January 17 in the year AD 807? In whose backyard did it stand for weeks or months before evaporating? We can also examine a simple block of wood and, through it, understand something of mass, dimension, extension, volume, presence, surface, exterior, direction, density, hardness, duration, flexibility, location, temperature, truth, life and growth. Such is the world of implications that is welling up within us as we extrapolate our understanding, go beyond what is immediate, and relate with what is outside of our reach.

What this all suggests is that it is not of crucial importance just how many things we come to know or how many places we happen to visit during our lifetime. We need not engage in frenzied pursuits of experience for its own sake. We need not to be troubled over our inability to be everywhere, to personally see and examine every place, person, and thing throughout the universe. By studying the few and understanding their richness, we can, by way of extrapolation, fathom very much about the many. Television can assist greatly in this endeavor.

The things of this world point toward endlessness in yet another sense that has little to do with numerical quantity or physical extension and everything to do with the far-reaching implications of existence itself. The fundamental appeal of the physical universe resides in the fact that any thing whatsoever—galaxy, sewing machine, horseshoe, eggplant, air bubble traveling upward from a turtle swimming underwater—is so vastly different from nothingness as to evoke everlasting wonder. The Book of Genesis implies that any transition from non-existence to existence—i.e., the mystery of creation—is an infinite leap. Even the existence of a tiny speck of matter, then, is ontologically expressive of an infinite journey that rests in the infinity of God. Therefore, we can meaningfully look at anything anywhere and exclaim with unmitigated appreciation, "You've come a long, long way!"

E. R. Goodenough of Yale University was concerned with such considerations of the infinite, and he wrote of the "*tremendum*" in our conscious and unconscious modes of awareness. Goodenough's term refers to the colossal continuum of endless dimension we experience on all levels; we must necessarily respond to it, even if we attempt to shield ourselves from it, suppressing it to some extent lest it prove overwhelming. All of us, though we may not consciously give it much thought, have some sense of boundlessness that forms the backdrop to all our thoughts and perceptions. Certain unusual things, such as the silent vastness of the sky on a clear night, astound us and electrify our sense of the mysterious in creation. This is a common denominator to all humans. One need not be highly educated to correctly sense the "*mysterium tremendum*."

All in all, everything in the cosmos harbors a mystery expressive of the endless. Within ourselves, answers to questions beget new questions, and the more we know, the more we become aware of how little we know.

It is not the function of science to provide us only with answers, as if to say that answers alone have meaning. Questions, too, are of high importance. They heighten our awareness of the endless, reminding us of the necessity of response to ongoing mystery. Science focuses on paying close attention to the ways of the world, which are truly "the ways of the wondrous" for one who is open to wonder.

To know a thing is to bless it with our recognition, a concept we can appreciate once we believe ourselves to be so empowered to bless. In knowing the world, we enshrine or dignify it in the realm of the endless that is ongoing within ourselves. This includes, but is not limited to, the "endlessness of lingering" wherein memories of our surroundings remain with us, ready to be recalled and reviewed for years and even for decades.

Memory enables us, although imperfectly, to recall and review our previously stored awarenesses. Value judgment comes into play here, as we purposely bring back to mind certain things to the exclusion of others. For we think of certain things as worth recalling, and others as deserving to be ignored or forgotten.

Imagination, too, comes into the picture. It enables us to juxtapose, compare, and relate to the objects of our knowledge. The resulting products of our imagination are like newborn creatures within the endless extensions of our mind. We pay attention to them and see them as alive with meaning.

From the standpoint of religion, our encounters with the world are of the utmost importance because they comprise the groundwork for our knowing God by way of creation. Assuming the immortality of the soul, there are numerous good reasons to believe that we are presently weaving a tapestry of a sort, a tapestry on which we will somehow gaze forever. Our afterlife promises to be a domain in which our recognition of the endless in this world and the eternity[2] of the next coalesce as one.

Ultimately, then, not only will we bless the creatures of this world for the privilege of our having known them, but we will know them, too, as having blessed us. In being what they were, they enabled us to know the Wondrous by way of the endless that was at work within us as we addressed this world with meanings in mind.

Each and every creature represents a specific actualization brought forth from countless possibilities residing in the mind of God. In this sense, then, every creature—including a broken spoke on a bicycle wheel—is a chosen one, displaying before us a specific coming-into-being in the realm of endless possibilities corresponding to the infinity of the Creator.

Notes

1. Joseph D. Hassett, Robert A. Mitchell, and J. Donald Monan, *The Philosophy of Human Knowing*, Newman Press (1953), 81.

2. Karl Rahner visualized eternity not as "time unending," but rather as the "finalized mode of spiritual freedom that was exercised in time." (See the final chapter in this book, entitled "Convergence toward Omega.")

For Discussion

1. Describe a past mistake of yours in terms of an "erroneous extrapolation" that was made on your part.

2. Do you consider yourself to be a "chosen one" of God? What leads you to think of yourself that way?

3. Can you imagine yourself knowing to full capacity, that is, being so full of knowledge that you can know nothing more?

CHAPTER 5

Rescue from Nothingness

"We have sinned and grown old, and our Father is younger than we."
—G. K. Chesterton

One of several landmark experiences I had during my early twenties was the reading of Gilbert K. Chesterton's *Orthodoxy*. His chapter entitled "The Ethics of Elfland," which described creation as marvelously extraordinary, especially resonated with me, for I had had similar feelings about the stuff of the world. He focused on Robinson Crusoe's shipwreck, and he viewed Crusoe's inventory of recovered items following the catastrophe as his "greatest of poems." That is to say, every kitchen utensil, every broom, ax handle, sail, hammer, rope, keg, lantern, and knife had become precious because Crusoe might have dropped them into the sea. Chesterton went on to write, by way of analogy, that everything everywhere has been saved from a gargantuan catastrophe—the "shipwreck" of nonexistence. Simply in being what it is, everything in the world is a great might-not-have-been that has been rescued from non-existence.

Along the lines of Chesterton's reflections, it becomes apparent that the transition from nothingness to any form of existence, regardless of how lowly it might seem to be, represents an infinite journey. For there is no such thing as an almost-but-not-existing creature—there are no intermediate steps between being and not-being. And, thus, (on a higher plane of thinking) infinity is at issue in even our most casual considerations of creatures. Ultimately, their infinite journey is enfolded or rests in God's infinity.

Chesterton lived in a kind of joyful delirium over the fact that nothing that stood before him had been "overlooked." He wrote of the fairytale philosopher who is glad that the leaf is green precisely because it might have been scarlet. He delighted in strangeness, and wrote in veiled humor that although rhinoceroses do in fact exist they look as if they do not.

GILBERT K. CHESTERTON
1874–1936

G. K. Chesterton was an influential literary figure in England during the first third of the twentieth century. He was known for his wit and wisdom, writing often in paradoxical and mildly satirical styles. An example of this was his statement that the trouble with worldly people is that they do not understand even the world.

A convert to Christianity, Chesterton maintained a close friendship with professed atheist George Bernard Shaw, and the two of them often joked about each other. Two of Chesterton's best-known writings are his *Orthodoxy* and *The Everlasting Man*. He also wrote adventure stories and poetry.

Personally, I was captivated by Chesterton's talent for seeing the new in the old, for regarding repetition in nature as a kind of ongoing theatrical encore. He insisted that sunrises and moonrises ought not to be taken for granted. In a mystical vein, he wrote:

Perhaps God is strong enough to exult in monotony [whereas we
are not]. It is possible that God says every morning, "Do it
again" to the sun; and every evening, "Do it again" to the moon.
It may not be automatic necessity that makes all daisies alike; it
may be that God makes every daisy separately and has never
grown tired of making them. It may be that he has the eternal
appetite of infancy; for we have sinned and grown old, and our
Father is younger than we.[1]

Chesterton noted that Christianity is like a sunken ship converted
into a submarine. After repeatedly being put down throughout the ages,
it has consistently resurfaced to change an apparently meaningless world
into something of significance and purpose. He delighted in the fact that
people are separate from things and are staring at them in astonishment.
He wrote about these things in 1908. Since those days the cosmos has
been revealing further truths about itself. These truths imply that we, as
well as things, might not be as separate from one another as Chesterton
had assumed.

In staring with astonishment at creatures today, physicists who are
concerned with exactness of measurement describe the world mathematically in terms of "quantum mechanics." Pushing analysis to its limits,
they find that atoms become "fuzzier" the more we insist on exactness in
measuring them. Out of this have come striking new insights.

In contrast with Chesterton's "separate from things and staring at
them with astonishment" statement, a more recent assertion by physicist
John A. Wheeler merits attention:

Nothing is more important about the quantum principle than
this, that it destroys the concept of the world as "sitting out there"
with the observers separated from it.... [When observing the
electron, for example] it is up to [the observer] to decide whether
to measure the position or the momentum....The measurement
[itself] changes the state of the electron. The universe will never
be the same. To describe what has happened, one has to cross out
that old word "observer" and put in its place "participator." [2]

Wheeler insisted that it is erroneous to visualize nature as an entity
that is apart from ourselves. So deeply are we enmeshed in nature that
our very observations of the world can effect changes in the world.[3] This
fact, Wheeler believed, must be taken into account if our knowledge of
the world is to correspond as closely as possible to the way it actually is.

We must reach not only outward, but also inward. As participating
observers of the cosmos, we ourselves are coupled to the whole of nature

and cannot dismiss any part of nature without also dismissing something of ourselves.

Our bodies are part of the cosmos. We are made of its matter—its water, its carbon, its electrons, protons, and neutrons. We have mass and we experience inertia. The electricity within our nerve fibers, for example, is part of the electricity of the cosmos. The atoms in our bodies, like those in the substances that chemists study, are held together by bondings. The core of each of our atoms, and also of those in materials apart from ourselves, is held together by mysterious nuclear forces. The breath of our life depends on the atmosphere that provides us with oxygen atoms. Yet we do not own our atoms. We borrow them to use throughout our lives and surrender them at death. If, as Einstein maintained, celestial objects cause the space around them to become "warped," then presumably our atoms, too—each in its own realm—are warping to some degree the space around us and within us, sending out gravitational waves every time we make the slightest of motions. (Note: The existence of gravitational waves is almost a certainty, but has not yet been confirmed.)

In recent decades, we have increasingly heard from philosophers, theologians, and psychologists that "no one is an island." Asserting that we affect one another in everything we do, spiritual writers teach that there is no such thing as a private wrong that hurts no one else. Nor can there be a secret wonderful act that does not uplift the whole human race. Now physicists, too, are making similar statements in regard to the world. They are insisting that gravitational couplings between ourselves and everything else are such that no one can do anything without in some way physically affecting every other creature.[4] Physicist Paul Dirac once said, "Pick a flower on earth and you move the farthest star!" The phenomenon of which he spoke is analogous to that of two persons lying in a waterbed. Every time one person moves, ripples of one kind or another, be they ever so slight, are sent out that will in some way influence the other person.

All in all, contrary to what some people say about science and religion being incompatible, it remains clear that we can, indeed, validly affirm both. It does not have to be either one way or the other. We can understand the world poetically and religiously as Chesterton did, and also scientifically as Wheeler did. Understanding things from multiple viewpoints is among the greatest of our freedoms.

Notes

1. G. K. Chesterton, *Orthodoxy*, Dodd, Mead, and Company (1908), 60.

2. J. A. Wheeler, *The Physicist's Conception of Nature*, Benjamin (1977), 244.

3. A rather crude example of this idea might be the statement that one can never determine the exact maximum speed of an automobile by using its speedometer, for the speedometer itself requires energy from the car to drive its parts, thereby slowing down the car ever so slightly.

4. This implies that we are connected with others even when alone, and also connected with people who have died centuries ago. But how can that be? Atoms are so extremely small that in a single breath we inhale billions of trillions of them. Assuming that the atmosphere has evenly mixed across the ages, this would mean that in every breath we take, we inhale some of the atoms exhaled by every other person who has ever lived—except, of course, those from newborn babies living far away. (It can be presumed that atoms exhaled by newborn babies have not yet had time to migrate around the earth so as to evenly mix in with other atoms throughout the atmosphere.)

For Discussion

1. Do you visualize yourself as one who has been rescued from nothingness?

2. Can you visualize your atoms as having been "borrowed"? What happens to you upon surrendering them?

3. Name a certain "ridiculous bit of reality" that merits attention from you. Why?

CHAPTER 6

The Wondrous World
of Interconnectedness

Among the great services that physical scientists have rendered the human family has been that of sensitizing people to the fact that all things are interconnected. Beginning with Newton, we came to understand that every object in the universe attracts every other object with a force that depends on the mass of the objects and their distance from one another.

Then came the behavioral scientists, who taught that we simultaneously influence and are influenced by our surroundings. Both physical and behavioral scientists acknowledge that widespread interconnectedness is at work in the functioning of the world. Ours is a world in which object A cannot touch object B without B also touching A. Isaac Newton expressed this long ago when announcing his famous third law of motion: To every action there is an equal and opposite reaction.

A phenomenon known as mutual coupling comes to mind. Let us say that you jump into a canoe and thereby produce waves. Soon afterwards, your waves cause my nearby canoe to rock back and forth. The rocking motions of my canoe will in turn send out waves of their own which will "feed back" and influence your canoe, which, in turn, will send out more waves to influence my canoe. Endless such examples of mutual couplings can be cited in the areas of science and technology. They can also be found in other areas, such as the sociological and the psychological.

Psychologically, people relate to each another in different ways. In some cases they are "loosely" coupled with each other in casual association; in other cases they are "tightly" coupled, as in ongoing family relations. So, also, in the world of things.

Scientists speak of things being "loosely coupled" as distinguished from "tightly coupled." An interesting example of tight coupling in the world of electronics is the phenomenon of packet radio: Let us say that I want to communicate with you by radio in rapidly sent Morse code. Our transmitters and receivers are designed to operate in unison with each other. Each time my transmitter sends a letter of the alphabet to you, it pauses afterwards to await a response from your transmitter, confirming that the letter I sent to you has been received. On receiving your confirmation, my transmitter is enabled or "given permission" to transmit the next letter in the message—and so forth throughout an entire message! The overall outcome is that once my transmitter finishes sending my message to you, I can be absolutely certain that you have received it. But, mind you, this happens letter-by-letter with extreme rapidity, thousands of words being successfully sent and confirmed each minute.

In the realm of the psychological, we do not so much as open a door without the door enabling us to know ourselves as its opener. Mutual coupling is especially at work in the arena of interpersonal relationships, where we cannot help others without knowing ourselves as helpers of others. So it is in the workings of nature: We can do nothing for people or things without them in some way shaping us. This principle leads us into the world of interconnectedness between matter and spirit.

The cosmos as a whole exemplifies a colossal interacting infrastructure within which humans seem to play a special role. Matter and spirit interrelate within us as we navigate the world and delve deeply into the realm of meanings. Sooner or later it dawns upon us that certain of our ideas lean heavily on the behavior of matter. For example, it strikes us one day that our very concepts of up and down are rooted in gravitational considerations. Matter pulling on matter holds our planet together. Our usual concept of "down" signifies "toward the center of the planet." Likewise, "up" means "away from the center of the earth that is held together by gravitation."

Within all humans (but especially in persons of religious faith) the material world outdistances itself. Within us the world makes a breakthrough. It behaves in ways beyond those in which matter alone is known to behave, beyond what the textbooks of physics and chemistry say matter and energy do when interacting.

Within us are the dual marvels of spirit moving body and body affecting spirit. So numerous are the kinds of couplings between matter and spirit within us that we can observe and experience them from endless different perspectives. These include our basic beliefs, which determine our courses of action, which lead us on to refine or redefine our beliefs; our body chemistry, which affects our feelings, which then alter our chemistry to some degree; and our visualizations, which trigger determinations, which in turn move us toward actions that further promote visualizations. We constantly interpret matter in the context of spirit, and we experience spirit in the settings of matter, such as when observing a beautiful painting.

By visualizing the world from the standpoint of *fields*, or surrounding regions of influence, physicists attempt to account for interrelatedness among things. For example, a gravitational field is visualized as existing around the sun. It is an "attractive" field that accounts for the earth staying with the sun instead of flying off on a tangent from its orbit around the sun. Another example: electrons that are made to move back and forth along the antenna of a radio station affect the electromagnetic field pervading free space. These effects are propagated through space in the form of invisible fluctuations that we commonly think of as radio waves and that travel extremely fast. On reaching our antennas, they set in motion the loosely bound electrons within them, the final outcome being the voices and music that we hear through our radios.

Fields and waves are among the most essential concepts in a scientist's ways of visualizing interconnectedness. They enable us to describe ways in which things reach out, ways in which everything is influenced by everything else.

Is it possible for a person of religious faith to validly hold that God and matter have something in common? Can God and "world" be thought of as interconnected beyond the commonly understood idea of Creator "up in heaven" somehow affecting creatures "down below" on earth? Does God have a personal stake in this world? Is it possible for us to think in this manner without engaging in pantheism? Indeed it is, and Christians can do so with very good reasons.

It is a basic Christian belief that God assumed a human nature in Christ, in a pivotal event in history that has become known as the Incarnation. Christians understand God as having taken on a living body like our own, having become a living human in this world and thereby establishing a new order of being. Christians also believe that the world itself has taken on new dignity through Christ. Matter, energy, space, and time, by association with Christ, have been raised to a new dimension. It is here that we come upon the wonder of interconnectedness at its zenith— the worldliness of God, as well as the Godliness of the world in Christ.

For Discussion

1. With regard to the interrelatedness among things, how do you visualize yourself as related or not related to various things, such as mountain goats, discarded tin cans, church steeples, radio waves, and ice cream?

2. In what sense can it be correctly said that God is "worldly"?

3. Relative to couplings, name a few of the desirable and undesirable couplings you have experienced in your life.

CHAPTER 7

On the World as Truth-Expression

"For this I was born, and for this I came into the world, to testify to the truth.
Everyone who belongs to the truth listens to my voice."
—John 18:37

One day long ago each one of us as an infant became personally aware of the cosmos. We awoke to physical reality. In time we became aware of ourselves as knowers of the entity "world," and of the world as object of our knowing. The vastness of the cosmic outreach challenged and nourished our imagination, and its mystery provided the backdrop for every consideration, reflection, and speculation we made. Yet the discovery of the cosmos is not the greatest of discoveries we shall ever make.

In facing the world we grew to discern that truth was somehow already there, as if having awaited our arrival and receiving us into its realm. On maturing, we came to experience ourselves as continually recognizing truths that are rooted in everything everywhere around us. So overwhelming is truth that even a false statement bears the trademark of truth to the extent that the truth of its untruth is often uncovered. So intrinsic a consideration is this that *we find it impossible to even imagine* a viable world in which truth would not be of utmost concern.

Each of us in some way unfurls and flies certain banners of our beliefs in highly visible personal displays that tell the world how we relate with its truths. No one is exempt from displaying these banners, even though we may not give them much conscious thought. For example, by wearing a wristwatch a woman displays her concern for knowing the time. Or, again, by neglecting to carry a watch, she proclaims another kind of creed, such as her belief in the relative unimportance of knowing the time with accuracy. Now, if she argues that she adheres to no creed whatever, she would simply be revealing her belief in a creed of non-belief. And what's more, even if she were to choose *not* to respond when asked what she believes, then her silence itself would convey the message that non-response is part of her belief.

Humans are by nature constrained toward external manifestations of themselves that reveal at least some of their inner and outer truth-beliefs. Our cosmic condition itself carries us into the beyonds of life, where genuineness is the ruling factor. Assuming that we behave in truthful ways, those who observe us cannot but have some sense of what we understand to be real or worthwhile in this world.

Whoever enjoys philosophizing about physical reality and our relationship with it eventually comes to understand that the concepts of truth, existence, reality, and being are one and the same. They form the baseline of what we address in our innermost awareness of the world. Concern for *that which is* is what we take for granted in science, religion and everything else of life … unless, of course, we choose to suppress the truth, a practice that leads to disappointment, rage, and perhaps even to disaster.

Truth properly recognized in one place, or in a certain discipline, cannot clash with truth in another place or area of thought because truth is expressive of that which is. This means that truths understood as scientific cannot falsify those understood as religious—which is not, however, to dismiss the existence of seeming paradoxes, which are due to the limitations of our minds and our language.

In paying attention to the countless substances and situations of the world, we not only extract what is true about things and events but also get in touch with something of supremely greater importance than the mere accumulation of facts. Sooner or later we move beyond our knowledge of individual facts and are drawn toward an understanding of the totality of truths as expressive of a living, life-promoting, all-pervading Fullness of Truth, an entity so supreme as to be the very Ground for any attempt to deny it. That is to say, we cannot deny our belief in truth without at the same time professing (or implying) in some way the truth of our denial. So fully immersed are we in this Universal Truth that we cannot dismiss it or circumvent it without wounding ourselves.

This Fullness of Truth is known today by different people of different cultures by various names, including Yahweh, Allah, Ultimate Reality, the Existent One, God, the Way, Holy Presence, Father in Heaven, Love Itself, the Alpha and Omega, and many other such titles.

Hopefully, there comes a day when each of us realizes how privileged we are to be in this world, to engage it, to utilize it, to investigate it, to avoid its greatest dangers, to experience with grace the passing of its time. Eventually we come to recognize the cosmos as the milieu of universal Truth-expression. We see it as the down-to-earth revelation of the Wondrous in the context of matter, energy, space, and time. From our experience of confinement within this world, we gradually awaken to see that this plight itself is so wondrous as to suggest that a Loving Concern is at work within the cosmos at large.

To recognize the loveliness of truth and to see that truth and love go together as one—this is, indeed, an enormous perception. Whoever enjoys this insight also understands that the world of "plain old matter" harbors an awesome promise of fulfillment. Such a person understands that nature itself leads us to the brink of what lies beyond this world. To discern the ultimates in this world, to recognize a quality of other-worldness in what we commonly call "world"—this is truly the zenith of awareness. And it can be ours at any moment, anywhere, provided we remain prayerfully receptive to it.

For Discussion

1. Is it possible to have a reality in this world that is in no way expressive of truth?

2. Should the personal beliefs of an ordinary citizen in the street be of concern to you?

3. Would your answer be the same if you learned that that person is a terrorist?

CHAPTER 8

The World Tells Us Who We Are

"Everything points to humans, as far as we are concerned,
as the key to the laws of nature, for there matter realizes its true potential."
—Adam Ford

I n a world where we create our own identities within the limitations of our talents, the subject of the relationship between science and the human spirit is critical. Consider, for example, our astrophysicists turning their dish antennas toward the skies. Some of the radiations they receive are believed to have originated from the primordial universe. The observations made by these scientists lead us to reflect on who we are in the context of the cosmos as a whole. And this we do while understanding ourselves as rooted in a world that behaves today in basically the same physical and biological ways (inclusive of mutations) as it has always behaved. Remaining conscious of this, we increasingly come to see ourselves—matter and spirit—in harmony with our perceptions of both science and religion.

Attempting to integrate science and religion to some degree, some writers insist that the Book of Genesis and the modern cosmologist's Big Bang theory (explained on page 126) are becoming increasingly harmonious. These writers caution, however, that lingering suspicions between scientists and theologians still exist.

How do we harmonize our understanding of the sciences with our ideas on philosophy, religion, and psychology? The question itself is enormous. The answers are many, and the pathway toward good answers is fraught with uncertainties. We can begin by searching back into our childhood, for the practice of reviewing our early experiences is a good way to gain new insights.

Recalling my own childhood experiments, I can remember my brothers and I using a hammer to pound various things. By smashing bits of glass and lead BB shots, we came to understand brittleness and malleability. We also tried to trick crayfish into coming out of their holes in the ground in broad daylight by covering the entrances of the holes with boxes. We thought that they might "conclude" that it was nighttime and, in acordance with their usual nocturnal behavior, come forth for food. The results were just the opposite. They stayed in their holes.

Also, on observing how worker bees visited flowers when gathering nectar, I once placed individual drops of honey on clover flowers, one drop per flower, in order to see how the bees would react. Some seemed to ignore the honey whereas others, on discovering the honey, remained on the flowers for fifteen minutes or more "until their tanks were filled," as I reasoned. Humorous as they seem in retrospect, these and other childhood experiments were resounding successes for me in the early 1930s.

Some adults would likely dismiss simple activities such as these as nothing more than child's play. However, we ought not to view childhood experiments as having low significance. Aesthetically speaking, childhood experiments fully qualify as meaningful engagements with the physical world, at least on the beginning levels of what certain psychologists and theologians call the "I–it" encounter. In these engagements we interact with the substance and situation of the world in very intimate ways, thereby learning who we are in relation to what is outside and apart from ourselves.

German theologian Karl Rahner (1904–1984) was attuned to this type of understanding. Unlike science-oriented persons who describe matter as "concentrated energy" or as "that which exhibits inertia," Rahner referred to matter in more human-specific terms as the *condition* that makes possible *the objective other*, which is what the world is to humans. He wrote:

> Matter means the condition for that otherness which estranges man from himself, and precisely in doing so, brings him to himself.[1]

He thus implied that the world has a parent-like quality. Its substance provides us our bodies through which we grow cognitively while searching deeply into our psychological "withins." Yet the world's substance does not allow us to become totally self-centered. It draws us out of ourselves, as when we necessarily reach out for food and drink. It compels us to come to grips with reality at large, thereby discovering who we are in terms of our inner cravings addressing our outer needs.

While visiting France, I learned an expression often used there. Whereas we might say, "That means...," the French say, "*Il veut dire,*" the literal translation being: "It wants to say." On reflection, it becomes

KARL RAHNER, S.J.
1904–1984

German theologian Karl Rahner is regarded by many as the foremost Roman Catholic theologian of the twentieth century. His doctoral thesis had to do with a new interpretation of the philosophy of Thomas Aquinas, and his influence grew following the early 1960s when he served as an official papal theologian at the Second Vatican Council. Rahner's theology takes into account evolution, existentialism and personalism.

The central theme of Rahner's thought is that human experience is without meaning unless it is seen in the light of God's mystery. Rahner demonstrated a willingness to address the concept of religious pluralism, taking into account the "anonymous Christian," such as the person who has never heard the Christian message but who in effect lives the Christian faith through a lifestyle of selfless love for others.

Rahner treated evolution as an open question. While avoiding actual endorsement of evolution, he proceeded to sketch a system of theology that would allow for Pierre Teilhard de Chardin's evolutionary outlook as well as for other viewpoints not opposed to the fundamentals of Christian faith.

apparent that this expression is not inappropriate for use in referring to a way in which we learn about ourselves.

As regards our physical capabilities, the world strongly reveals to us who we are. It wants to say, for example, "You were never meant to be a pole-vaulter." Or, "You display great skills when cooking biscuits." By way of value perceptions on our part, it might induce us to declare, "Swimming the Amazon River is not worth the risks!" Thus, by matching ourselves against the world, we fine-tune our understanding of who we are, for the world unremittingly reveals things about itself, and about ourselves in relation to it. What we learn from our surroundings strongly determines the extent to which we know our places in this world.

But when we then proceed to organize our gathered knowledge into various fields of thought, we risk segmentation and a loss of wholeness. Therefore, it might be of value here to cite specific instances where diverse fields of thought relate closely with each other. An example involving science, religion, psychology, and movie production comes to mind:

Science experiments reveal that small, lightweight objects—the reeds of an oboe, for example—vibrate very rapidly. They emit high pitched sounds in the process. Larger, more massive things, possessing more inertia, exhibit greater slugishness of motion when vibrating freely. They vibrate less rapidly and produce lower pitched sounds. The tiny inertia of a child's vocal chords accounts for the child's voice being of higher pitch than the voices of adults whose vocal cords are more massive.

In the world of Hollywood moviemaking during the mid-1950s, Cecil B. DeMille produced a film entitled *The Ten Commandments*. Among its most moving scenes was the one in which the prophet Moses (played by Charlton Heston) received the commandments from the Lord.

Let us ask ourselves how, if we had been in DeMille's shoes, we would have portrayed God uttering the commandments atop Mount Sinai. Certainly, we would not have used a high-pitched canary-like voice to portray the utterances of the Lord. Such a voice would lack power. It would not be able to convey the weight of seriousness required by the scene.

DeMille utilized an extremely low-pitched voice, possibly from a bass voice being recorded and then played back at half speed or slower. He used extremely low-pitched sounds in order to touch the psyche of his audience. Recognizing that people associate low-pitched sounds with massive objects, he used that knowledge to verbally dramatize a message delivered by God. The low-pitched sounds were intended to suggest God's immensity. Or, stated less formally, by vibrating matter rather slowly, he obtained the sounds that would bring his audiences to visualize what religious writers refer to as God's overwhelming immensity and all-pervading presence. Vibrating matter being used to portray awareness of God! How strange it is!

Our relationships with matter are, indeed, often very intimate. It might be the color of the paint on a particular barn. A certain toy saved from childhood. A favorite pair of earrings. A photograph of a deceased grandparent. An antique rocking chair, stripped of its old varnish and then re-varnished. Things such as these cast their charismatic shadows over us and have much to do with who we know ourselves to be. Through the attention given them, we come to know ourselves as recognizers of their uniqueness, as devotees of their charm, as persons who appreciate both the old and the new, who are willing to go into action on their behalf (such as when we move a wine glass away from the very edge of the table on which it has been placed).

The presence of things throughout the cosmos enables us to discover and explore transcendental qualities within ourselves. Each of us possesses the ability to recognize the worth in creatures despite their being ignored, discarded, or rejected by others. In this sense we become their liberators, the ones who give attention to the supposed insignificants of creation and crown them with meaning.

It is we who unite the segments of the world in meaningful ways, relating things with things—fish hooks with fish, coat hangers with coats, carpets with floors, hammers with nails. We elevate things everywhere in countless ways by recognizing, affiliating, devising, directing, and transforming them. By giving particular things our close attention and endorsing their potential, we subconsciously bless them, save them from nullity, and implicitly deem them to be worth our recognition as more than insignificant "mere stuff."

It is in this sense that we spiritualize the world, elevating matter, energy, space, and time in the realm of value recognition. Through humans, the cosmos itself becomes a new creation in both the physical and the spiritual sense. And, as if in return, the cosmos reveals to us the countless ways in which our presence constitutes an enhancement within it.

Most of us are familiar with the concept of "defining moment" or "moment of truth." We generally use those expressions when citing incidents where a challenge must be faced, such as when a bullfighter confronts a bull. However, we seldom think of our ordinary engagements with simple things in terms of defining moments.

I was walking along a sidewalk at 7:42 on a Thursday morning. A baby frog hopped across my path, and I hardly noticed it. Suddenly a new and powerful awareness hit me: "Do not dismiss that frog so quickly!" I told myself. And so, I reconsidered the frog, translating my previous understanding into higher terms: "Look, friend," I told myself,

"Don't underestimate what you just witnessed! Whether or not you happen to like frogs, understand that from now on you will forever know yourself as a person who once saw a tiny frog crossing his path at 7:42 on a Thursday morning. And, from this moment on, when someone asks who you are, part of your answer can be, 'I am one who witnessed a tiny frog crossing my path at 7:42 on a Thursday morning.' Be aware that that frog encounter will stand as part of your ongoing definition, as a slice of your very personal 'I am' assertion for the remainder of your life. And also beyond! Indeed, it's an exceedingly tiny sliver of your total definition, but nevertheless it is a real and genuine part. It can certainly be forgotten or dismissed, but it can *never be erased or undone*! If you were to repeat that 'I am one who witnessed a tiny frog...' assertion thirty years from now, it would still hold true. It stands forever!"

A further realization then came to me: If this understanding holds true in relation to a frog that crossed my path, then it holds true also in my relationship with the sidewalk that was crossed by the frog, with the fence that paralleled the sidewalk, with the white clouds passing overhead, and with every blade of grass along my way. In fact, to be consistent I must acknowledge that everything in the whole of the cosmos contributes its part toward defining who I am.[2] The encounters that we have had with every thing everywhere at every moment along our way can never be nullified, for time cannot be reversed such as to erase what has already taken place. In this sense, at least, a lingering characteristic of permanence resides in the worldly definitions of who we are—this being an innate attribute of the cosmic mystique.

Taken alone or together, then, the endlessly abundant statements that add up to say who we are all participate in forming us, *even when ignored or long forgotten by us*. Every moment of our existence is in some small way a defining moment, and in the ongoing progression of time that gives continuity to these defining moments, they all add up to the fabulous continuously flowing entity that we call "experience of life."

Furthermore, it is not only the presence of things that gives definition to each of us, but also their absences. Consider a shadow, for example. Shadows are recognized patterns formed by the absence of light. They exemplify negation, exclusion, and lack of presence. And among our "I am" statements, we can define ourselves at least in part as individuals whose shadow, or configured absence of light, follows us wherever we go while exposed to light.

Ultimately, though, with regard to both the physical and the spiritual worlds, we must consider the infinite when searching for the full disclosure of who we are. We must keep in mind that the *totality* of who we are does not rest solely in the summation of what each piece of the material world tells us about ourselves. There is more.

In addition to the material world defining who we are, there exists another realm at work in defining us—namely, the non-material or spiritual world that tells us who we are in terms of the good that we choose and the evils that we reject, for humans can also be defined in ethereal terms of love, appreciation, devotion, patience and other qualities that clearly extend beyond the physical worlds of atoms and molecules.

Notes

1. Karl Rahner, S.J., *Foundations of Christian Faith*, trans. William V. Dych, S.J., Seabury Press (1978), 183.

2. It is this writer's opinion that every atom within the makeup of the cosmos, simply by existing, contributes its distinctive part toward defining who we are, the ongoing awareness of this fact being a very exciting experience! Obviously, we can speak intelligently about things that we have never personally encountered. For example, the statement: "I am a person who has never seen an igloo," expresses the "I am...who" of someone in terms of a negation. Most of the cosmos lies beyond our personal observation and experience. Yet, all of it—every atom—contributes something at least negatively toward defining who we are.

For Discussion

1. Recall a childhood experiment of yours and tell about its outcome.

2. Define yourself in terms of certain material things that you like and do not like.

CHAPTER 9

Cosmic Enhancers

"The color we see is not in the world around us. The color is in our heads. . . .
How nice that [our] eye-brain interactions produce the beautiful colors we see."
—Physicist Paul G. Hewitt

In order to present this topic more effectively, I have chosen to focus on the college classroom, where across the years I have tried to make my students more aware of who they are. Repeatedly I suggested that, when relating with the world at large, we are constantly acquiring new perceptions of ourselves that we repeatedly update and extend. I often reminded them that the sciences offer us unique opportunities to know and understand who we are in relation to the world.

I sometimes digressed a little from the physics of things and explained to the students how their senses function in relation to the outside world. "After all," I reminded them, "Scientists utilize their senses when making observations of the world, and, thus, it is wholly proper for us to focus for a moment on our senses!"

"You know," I would continue, "Things out there apart from us are *not* hot and cold in our customary ways of thinking of them as having different temperatures. Our experiences of temperatures correspond to faster-vibrating or slower-vibrating molecules acting violently against our skin. Bear in mind that there is no such thing as a hot or cold molecule! It is the accumulated *motions* of molecules that we sense as temperature. Individual molecules have no temperature. As molecules are vibrating faster in a material with which we are in contact, we sense the material to be 'getting hotter.' Why? Because the faster-moving molecules of the material, when bumping against our skin, cause the molecules in our skin to vibrate faster than before. This causes our touch sensors to generate electro-chemical pulses which, after traveling through our nerves, reach our brain and arouse within us what we experience as sensations of heat. Conversely, slow-moving molecules instigate in us feelings of cold."

I enjoyed speaking to them in this manner also when teaching about sound: "Take sound, for example. What we call 'sounds of music' are *not* musical tones that travel through the air as loud or faint in their intensities, or as high and low in pitch. What travels through the air are compression waves of various frequencies and intensities, pressure variations of air molecules that are bunched together. Nothing about them is musical at this stage. But on reaching your eardrums, they stimulate your hearing sensors, which in turn generate electrical pulsations that travel along your nerves toward your brain. On entering your brain they are translated at last into the inner experience of yours that you call music."

Extending the examples into the world of light, I continued: "Most persons think erroneously of light as something of color and brightness outside of themselves, something that radiates from things such as a camp fire and enters their eyes after traveling through air. This is certainly not accurate. Light as light, with its hues of color, does not exist that way outside your brain. What enter your eyes are electromagnetic waves of various wavelengths and intensities. On reaching your retinas, the waves are converted into electrical pulses of various kinds. Traveling through your optic nerves, the pulses reach your brain, where they are translated into the awareness-experiences of yours that you recognize as light of different colors and brightnesses. That is to say, color and brightness arise *within you* as personal (subjective) experiences in the final steps of this process. The incoming radiations from rainbows and sunsets exist as color only in the brains of their beholders, and not 'out there in the sky' where they seem to be.[1]

"The process is similar for your sense of smell," I continued. "On entering your nose, different vapors trigger your olfactory sensors. These, in turn, produce electrical pulses of certain frequencies that travel to the part of your brain that excites sensations of smell. The more intense vapors generate higher-frequency electrical pulses that your brain translates into the experience of 'smelling stronger odors.'

"Now, relative to different 'flavors' of odors, these arise from different segments in a localized region of the brain responding differently to different kinds of vapors. Thus, within the brain electrical pulses are translated into personal experiences of odors. What we call odors do not exist as odors apart from ourselves!"

I concluded by saying to the students, "Thus, it is to the *inner world* of ours in the realms of sight, hearing, smell, taste, and feeling that we respond while addressing the world outside of ourselves. This in no way diminishes or invalidates the value of our contact with the external world, for we learn early in life that a one-to-one correspondence exists between *what is actually* 'out there' and *what we perceive* as 'out there.'"

There is very much more that could be said about the aesthetics of "world," about our responses to meanings that we recognize in what our senses deliver to us. On hearing an orchestra, for example, we immediately transcend the mere mechanics of the incoming waves, finally speaking of the sounds we experience, using words of aesthetic expression such as beautiful, depressing, bouncy, or jazzy. Similar evaluative statements could also be made with regard to the responses of our other four senses.

Observation being an essential part of their engagements, scientists do well to mull over what is taking place within themselves when they explore the world, for they themselves are a major component of "world." Overwhelming evidence points to the fact that no one learns about external things or events through a knowledge that is immediate. An interesting point in this regard was made by the late scholar-psychiatrist, Carl Jung, who wrote:

> As a matter of fact, the only form of existence we know immediately is psychic. We might well say...that physical existence is merely an inference, since we know matter only insofar as we perceive psychic images transmitted [to our brain] by our senses.[2]

Jung was insisting that our knowledge of everything in the physical universe is indirect. Even when feeling things with our hands, our knowledge of those things is indirect. Stimuli derived from things and events must undergo conversion into electrochemical signals that travel along our nerves and reach our brain before the things can trigger our awareness and become known by us. They must become *experienced psychic phenomena within our brain* before we can relate with them as being outside of ourselves. Because of this, Jung believed that internal psychic phenomena are as real to us as external physical phenomena, and perhaps much more so.

"The world at large must be bleak and dismal without people!" one of my students remarked.

"Definitely so!" I responded. "However, we ought not to think of ourselves as deprived in any way by nature when nature insists that we translate what it presents to our senses. Quite the opposite! It would be fully proper for us to visualize ourselves as favored by nature, which escorts us on journeys of enrichment. Think of yourself as an enhancer or 'glamorizer' of nature!" I suggested.

On hearing this, one young man replied in a spirit of fun: "You mean to say that if scorpions or alligators experience colors, then it means that they, too, are enhancers of nature?"

"Why, yes! In ways that are proper to themselves, yes!" I answered. "Were it ever confirmed that they do possess color vision, would that disturb you? And, if so, why?" I teased. "If we visualize God as a multiplier of favors, should it disturb us to learn that an animal or an insect which we might hold in disdain enjoys in some way what we ourselves enjoy?"

Everyone laughed, and then the young man posed a further question: "What's the good of knowing all of this stuff about how we come to sense things in different ways?"

"It's a matter of richness of experience," I answered. "Our desire for delight in what we experience acts as a driving force that lifts up our spirits, drawing us upward into the world of wonder. Let me give you an example: It is one experience to turn on a radio and listen to a faraway station. However, it's a much richer experience to understand how a radio works, how our ears and our brains work to enable us to hear music. The more we understand the processes, the more we are able to appreciate the outcomes. Whereas most persons might appreciate music, you yourself can not only appreciate the music but also delight in understanding the physiological process of music recognition while it is taking place within you. *Appreciation* and *richness of experience* go together. Our understanding of science enable us to see things with greater depth, to visualize in expanded dimensions what stands before us," I concluded.

Nature impels us toward ceaseless new awareness as our sensing organs receive and translate the information arriving from our surroundings. The enriched inner experiences that result lead us to new thresholds of wonder. Within our neurological system the world is lifted up as we rise above "the mere hardware" of things and speak of them in terms of their workings, their qualities, their appeal, their beauty, and their truth. When asking ourselves who we are, factors such as these ought always to be considered.

When pursued at length, our relationship with matter-energy and space-time takes us on far-reaching journeys of self-identity. From the viewpoint of enhancement, a two-way involvement is at work in the world. Not only do we enhance nature, but nature outside of ourselves also enhances us. A teacher friend of mine comes to mind.

Shortly before his death in 1967, Charles Hall related an experience he had had as a young construction worker in Rochester, New York. As he was working along the outer wall of a tall building, the boom of a massive swinging crane caught him by surprise and forced him against the wall. The boom's tip passed about an inch from his stomach. He explained how he afterwards interpreted the experience in a deeply personal way. He understood it as a message delivered to him through the material and workings of that situation—the huge crane, its motion, its inertia, the tiny air space between himself and the boom. The close encounter jolted him into a new understanding of who he was. He realized that he had been spared *for a reason.*

The meanings he perceived through that experience marked a major turning point in his life. Driven by a deep spirit of gratitude, he left everything and entered religious life. He spent his remaining years serving others, teaching science to disadvantaged high school boys until cancer claimed his life.

Notes

1. Although imperfect, there is an analogy here that is worth considering: Take the case where a faraway TV camera is focused on the American flag. Does the faraway red, white, and blue of the flag come to your house in the form of light? Indeed not! Rather, those colors are translated at the TV station into electromagnetic waves of different frequencies and pulse rates. These, in turn, are sent out as invisible waves that reach your home antenna and TV receiver. The receiver then translates parts of those invisible incoming waves back into the original visible red, white, and blue light that is finally displayed on its screen. In a poetical manner of speaking, then, one might say finally that your TV receiver "enhances" the invisible incoming waves into what is finally displayed as red, white, and blue.

2. Carl G. Jung, M.D., *Psychology and Religion*, Yale University Press (1938, 9th printing in 1955), 11.

For Discussion

1. Does the realization that the music you hear and the colors you see exist within your innermost self and not "out there" disturb you in any way?

2. Can you recall any particular event in your own life that dramatically changed your understanding of who you are?

CHAPTER 10

The Mysticism of Kepler and Newton

Among the most interesting examples of a scientist trying to unify almost everything was Austrian astronomer-mathematician Johannes Kepler (1571–1630). Kepler formulated his three laws of planetary motion after having tried for many years to figure out the shapes of the planetary orbits. Thinking of God as the architect of the universe, he reasoned that God must have utilized the laws of geometry in designing the shapes of the planetary orbits. So he regarded geometry as holding the key to discerning the mind of God as it related to the workings of the solar system.

In Kepler's eyes, astronomers were "the priests of God called to interpret the Book of Nature."[1] He observed that in the world of astronomy there were six (then-known) planets, whereas there are five, and only five, regular solids[2] in the world of geometry. He was intrigued by the idea that the regular solids might be made to fit snugly into the five spaces between the orbits of those six planets represented as thin-shelled spheres. He actually constructed a small scale-model in which spheres and geometrical solids fitted together quite snugly. The diameters of Kepler's spheres were made to relate with one another in the same proportion as the known diameters of the orbits of the planets.

Kepler even went so far as to attempt to relate the notes of the musical scale with his geometrical model of the solar system. He visualized the solar system as representing a huge celestial musical instrument.[3] As he imagined them, the larger planetary orbits would emit lower-pitched musical tones whereas the smaller orbits would emit tones of higher pitches. In effect, Kepler was attempting to interrelate the solar system, the world of geometry, and the mind of God with aesthetics of music. Indeed, an ambitious dream!

Kepler's musical model is referred to as "the music of the spheres" or "the harmony of the world." His writings describe the planets as emitting intelligible but inaudible melodic tones while they "hum along" in their orbits. In a mystical vein, he further visualized the choirs of people on earth to be imitating this planetary music, their singing itself having meaning only insofar as they chant in the realm of time that he called "the everlastingness of all created time in some part of an hour." What an imagination!

As chance would have it, in trying out various ways in which to fit the planetary orbits into the musical scale, Kepler discovered what came

JOHANNES KEPLER
1571–1630

As a university student, Johannes Kepler was entranced by the ideas of the astronomer Copernicus. Kepler went on to work with the elderly Danish astronomer Tycho Brahe, who died in 1601, leaving Kepler with his enormous amount of data collected on the positions of the planets at various times across many years. Using Tycho's data, Kepler finally formulated his three laws of planetary motion:

1. The planets move in elliptical orbits with the sun at one focus.
2. A line joining a planet to the sun sweeps out equal areas in equal time periods.
3. The mean distance of the planets from the sun, when cubed, is directly proportional to their orbital periods squared.

to be known as his third law of planetary motion.[4] This very important law was formulated in 1619 as a side comment, a footnote hidden among eccentric musical fantasies in his work entitled *Harmonice Mundi*.

As for his first law, a certain observation of the position of Mars made years before by Tycho Brahe finally convinced Kepler that the planetary orbits were elliptically shaped rather than circular or oval. This led poor Kepler to discard his glorious dream of putting everything together in a single vision, thus vastly modifying his idea of who God is.

Yet, in spite of all his setbacks, he remained very philosophical, and once wrote: "My friends, do not sentence me entirely to the treadmill of mathematical computations, but leave me time for philosophical speculations, which are my only delight" (*Gesammelte Werke*, vol. XVII, p. 79).

It is natural for us as human beings to try to interrelate people and things. Beginning in early childhood and continuing throughout life, we repeatedly attempt to integrate and simplify our visualizations of a highly complex world. Personally, I can remember trying, at about age six, to relate lollipops, soda pop, and automobile taillights. How? Well, I had noticed that, when they were lit, the red taillights of my father's automobile resembled the color of my red lollipops when the sun shone through them. So I asked my father whether the two substances—the red candy and the red taillight lens—were made of the same material. Then I attempted to produce a quantity of red strawberry soda by laboriously dissolving a red lollipop in a glass of water. Nothing much came from my experiment. At sundown I simply dismissed it. However, in later years I realized the importance of two lessons I had learned from it:

"Careful. Appearances can be deceiving!"

"Just because things have similarities does not mean their materials are the same."

These simple experiences illustrate a force at work within us from childhood, namely *transcendence*—the movement toward levels of awareness beyond what is immediate. Transcendence comes into play when we cease picturing our world as "plain old matter" and begin to recognize ourselves as privileged to be a part of it. It involves the use of our imagination and a willingness to navigate deeper in spirit. In the words of the late Thomas Merton:

As humans, we look within ourselves and discover an unspeakably deep dimension. Then we look outside and find depth without end there, too. And in time it grows apparent that the two worlds are one and the same.[5]

Somewhat like Kepler, Isaac Newton (1642–1727) also chose to view science in terms of God's power to produce a perfect creation.[6] Newton visualized God as having purposely specified the earth's mass, velocity, and distance from the sun so as to enable humans to survive and thrive.[7] He thought of God as having tilted the earth's axis relative to the plane of its orbit in order to provide seasons for humans, sparing us from the boredom of a year with only a single season. Thus, Newton's study of physics was also a reflection on God's immediate concern for humans.

In those days, when Europeans were moving beyond the use of the hourglass, clocks held a special fascination. The discoveries of Kepler were therefore perceived in terms of the so-called "clock model" of the solar system. In this view, God was pictured as the great Adjuster of the solar system, ever ready to speed up the planets if they began to slow down or to slow them down if they began to speed up, so that they would not come in "late" or "ahead of schedule" which, of course, they never did. Philosophers thought that it was *because of* God's ongoing adjustment activities that the planets consistently came in "on time."[8]

A noteworthy aspect of the mentality of these early scientists was their ongoing recognition of matter and spirit working together. They not only thought of God as First Cause of the universe but also published their scientific manuscripts with God foremost in mind, viewing God as the immediate and efficient cause of the physics they observed and recorded. It was a practice that continued into the early 1800s, as evidenced by Emperor Napoleon Bonaparte's complaint that a certain scientific manuscript contained no mention of God. Modern science, to a large extent, has put aside the ethereal idea of what Aristotle called "the final causes" of things, focusing instead on behaviors such as how things perform under different conditions.

Everyone everywhere, at least on personal and subtle levels of awareness, is caught up in the practice of relating matter and spirit. Indeed, we cannot avoid the practice, for we ourselves are "enspirited" matter. In recognizing and responding to each other, we put into action our deep-seated beliefs that matter and spirit do, indeed, interrelate and work together.

Granted, the varying interactions of the human mind with the world must be taken into account when we talk about observing the world. But in spite of individual differences, there are certain similarities. Like people of ancient times, we look both inward and outward throughout the whole of nature and experience wonder. Whereas they looked at mountain peaks, reindeer herds, and crops with wonder, we are dazzled today by those same things and also by cellular phones, computer communications, satellite weather pictures, global positioning systems and very much more.

SIR ISAAC NEWTON
(1642–1727)

The English physicist and mathematician Isaac Newton has been long regarded among the foremost scientific intellectuals of all time. He is best known for his universal law of gravitation, his work on light, and his development of differential and integral calculus. Although unable to understand how objects could attract each other at a distance without being physically connected, he was nevertheless able to accurately describe the manner in which gravitational forces depend on masses and distances. He was also able to explain why the moon remains in the heavens whereas apples on trees fall toward the earth. Here is the mathematical formula that expresses his law of gravitation: $F = G\, m_1 m_2 / d^2$.

Like those who went before us, we pursue meaning through the material, groping for ultimates in what is immediate, experiencing wonder in the ordinary, discerning something of the everlasting in the temporal. In addressing the stuff of the world within the context of spirit, we grow in awareness of who we are—and, also, of who God is.

Indeed, Johannes Kepler engaged in mysticism. And so did Isaac Newton, who spent some of his later years delving into theology.

Notes

1. Arthur Koestler, *The Watershed: A Biography of Johannes Kepler*, Doubleday (1960), 61.

2. A regular solid is a polyhedron the faces of which all have sides of equal length and equal angles between the sides. They include the cube, the tetrahedron, the octahedron, the dodecahedron, and the icosahedron.

3. His approach was not unlike that of Louis de Broglie fitting his "standing matter-wave patterns" around the orbits of Bohr atoms in 1924.

4. John Hollander, *The Untuning of the Sky*, Princeton University Press (1961), 41.

5. Paraphrase of Thomas Merton's words from a TV show about his life, June 1984.

6. Dietrich Schroeer, *Physics and Its Fifth Dimension: Society*, Addison-Wesley (1972), 95.

7. In comparison, a modern view might be that humankind "just happened" to develop on a planet that was at a suitable distance from the sun and had favorable conditions for the evolution of biological life as we now know it.

8. In comparison, a modern view might be that the planets come in "on time" because, given the conditions, there is nothing else that they can do.

For Discussion

1. Can you recall certain instances, especially in your early years, when you tried unsuccessfully to relate things with each other?

2. Relative to the phrase "God's ongoing adjustment activities," what kind of "adjustments," if any, do you visualize God as making in our everyday activities?

3. What do you think of the statement that the planets complete their orbits "exactly on schedule" because, given the conditions, there is nothing else that they can do? Is this an adequate explanation?

A NASA localized view of the surface of the planet Mars. Although not shown in color here, the surface actually has an orange-ish hue, which helps us to understand why Mars has traditionally been called "the red planet." How thrilled Galileo, Kepler, and Newton would have been to see this photograph!

CHAPTER 11

Oddities of Planet Earth

While adjusting ourselves to the ways of this world—its days and nights, its gravity, its rain, its air, its annual cycles—we sometimes fail to see that much of what we consider ordinary or mundane is in fact very unusual. In this chapter we will consider certain strange ways of our planet that might ordinarily escape our attention. To do this effectively we will need to imagine our earth from the perspective of visitors from outer space: persons who might have been born in a spaceship where zero-gravity has been an ongoing personal experience for them.

But first, let us recognize that designating something as "odd" is a very relative statement. When we call things odd, we are viewing them as exceptional, or strange, or out of step with our usual experience. However, there are instances when our ordinary experiences, as viewed from other perspectives, can also be understood as strange. In the adventurous spirit of this chapter, I will be giving a few examples of this strangeness from the viewpoint of physics.

At this point in human history, we have benefited from over three decades of zero-gravity experiences. Most of us have fairly good ideas of what life must be like in zero gravity environments (ZGEs). We have witnessed astronauts taking "walks" in space and floating inside their space stations. And now we can reverse our perspective and try to imagine the experience of intelligent creatures from a faraway ZGE coming to investigate our planet. What would their experience be like? What kind of stories about planet Earth would they tell on returning to their own world?

It is within this larger context of the universe that we acknowledge matter to be the exception rather than the rule. Actually, we might well view our whole earthly experience as an exception throughout the cosmos as a whole. While still sensing a certain strangeness about ZGEs "out there," we must remember that most of the space of the cosmos exhibits what, practically speaking, we would consider to be zero gravity, for gravitation is significantly sensed by humans only as we come closer to stars, or to moons, or to planets such as Earth.

The experience of gravity, of course, is ever-present to us as we navigate our planet, feeling our weight pressing against floors and against the chairs in which we sit. All of this happens while we breathe air that, fortunately, is—like ourselves—held against the earth by gravity's pull. Church bells ring, birds fly, rain falls, and coffee pours downward into our cups. These things are all familiar, so familiar, in fact, that we rarely think to question their behavior.

But what about life away from Earth, in faraway places or in the weightless mode of astronauts orbiting the planet? What would life be like where most of space is empty and lacking matter?

Walking as we know it on the earth would be impossible in a ZGE, with one possible exception. If an astronaut were in a spacecraft that happened to be whirling, then centripetal forces would arise. And if the spacecraft were spun at the correct speed, the astronaut would be pushed outward against the walls of the craft, giving him or her the ability to stand against the wall as if standing on the earth. This exemplifies the phenomenon of *artificial gravity* mentioned in physics books.

Dances performed in a ZGE would somewhat resemble the dances performed on earth by people in swimming pools. Arms could be whirled, feet could be touched, bodies could be twisted, heads could be turned. The body as a whole would be suspended. Dancers would be unable to dart about from one location to another as they do on earthly floors.

The dancers could artificially maintain contact with the floor of a spaceship by gripping stirrups or handles attached to these surfaces. But they would lack the friction that we commonly have between ourselves and our floors, and on which we depend for our acceleration in maneuvering from place to place on earth. The dancers might wear magnetic shoes to perform on a metallic stage, thus navigating from place to place along the stage floor. But a further difficulty arises from the well-known phenomenon described by Newton: To every action there is an equal and opposite reaction. That is to say, a group of dancers moving suddenly to the right would jolt the entire spaceship and its contents to the left. When they accelerated to the left, the ship and its contents would respond by moving to the right. This effect exists in theory here on Earth, but it remains unnoticed because of the huge mass of the earth compared to the small masses of people. In a spaceship the effect could be neutralized somewhat by pairing off dancers, such that translational movements of one dancer to the left would be offset by simultaneous movements of another to the right. What a challenging experience it would be for a choreographer to plan dances for entertainment in a ZGE!

Two important physical phenomena that must be reckoned with in a ZGE are *buoyancy* and *inertia*. As experienced on earth, the forces of buoyancy are generally upward, and they account for dirigibles being suspended in air and boats floating in water. Being gravity-derived, however, buoyancy itself disappears in a ZGE. This means that smoke would no longer rise but would rather travel outward, expanding in all directions from the source. A flame from a candle would have a spherical shape instead of the elongated shape we normally see. Such a spherically shaped flame, of course, would quickly melt the wax and rapidly destroy the candle. Ordinary cups would be of little use, as they would not effectively hold liquids unless equipped with covers.

Such gravity-dependent phenomena would be absent in a ZGE. However, the phenomenon of inertia, which is intrinsic to matter, would remain the same everywhere. Things, including people, bumping into each other would essentially be jolted as they are on the earth. If, for example, a game of American football were played on a flat field in outer space, the players would be jolted just as they are on Earth when colliding with each other. However, passes would be thrown in such a way as to make the ball travel in straight lines to the receivers. Never could a ball be thrown with an arc that would fly over the heads of defenders and then come back down to receivers.

The above examples illustrate a few factors to be reckoned with when changing from earth experience to zero gravity experience. But there are psychological factors as well:

If orbiting near the earth, an astronaut would experience one day as lasting only about ninety minutes. How might he or she speak of enjoying fair weather in an environment that is without weather? Also, there is no such thing as *up* and *down* in a ZGE. These words would probably be replaced by *outward* and *inward*, or perhaps by *toward* and *away from* some reference point.

Finally, and just for the fun of it, let us consider our earth experience from the viewpoint of a visitor who has never before experienced gravity. Let us imagine this visitor coming to our planet and then returning home and relating the experience to a friend as follows:

As you know, Tok, we have no such things as curtains here. Curtains are things on Earth that are made of cloth and attached to the inside walls of rooms. These extend in a direction they call "down," which means "toward the center of the earth." And, whereas we stretch our clothes for storage, they "hang" their clothes after attaching them to so-called "coat hangers." And the funny part about it is that this word *hang* finds its way into their psychology. They encourage each other to "hang loose"—whatever that means—when hard times come. And they speak of "hanging out" when they're sitting around in groups doing nothing in particular. It's bizarre out there, Tok.

And, Tok, it "rains" on Earth. Water comes traveling toward the earth from the outside, always "downward," as they say, never "upward." Seemingly, it travels downward without ever traveling upward. And, upon striking the earth, it forms into "puddles," which are unlike anything we have ever seen. The surfaces of their puddles are flat and parallel to the surface of the planet, but perpendicular to the directions of what they call "up" and "down."

Earth people travel in small mobile boxes, Tok, which are mounted on wheels. These they call "cars." When pulled by gravity against the surface of the earth, the wheels of a car interlock somewhat with the surface to create a force called "friction" that stops the car when a "brake pedal" is pushed. And they bore holes into the planet to extract a kind of juice that they use to reduce friction wherever they wish to. They complain about the cost of that juice "going sky-high"—whatever that means. Also, they heat the juice, condense its vapors, and burn that in their cars.

When people on earth wish to show special respect for you, they tilt their upper body forward somewhat. They call that "bowing." And, immediately afterward, they bring their heads back up against their pull of gravity. When they light a match, the flame stretches upward instead of bulging outward. The same happens with their smoke. Our smoke expands outward in every direction, but not theirs!

Also, on Earth there are animal creatures of many kinds. All of them are subject to gravity. But among their animals is one called a "cat" that seems to be partially exempt from the usual pull of gravity. If you hold a cat upside down and release it, gravity seems to pull its feet downward especially hard, and its feet hit the earth ahead of the rest of its body. People refer to that as landing "right side up." It's weird, Tok!

Whenever you throw something directly outward from the earth, it slows down, turns around, and comes back to you. Isn't that crazy? It's almost as if you have an invisible rubber rope tied to it. And yet they claim that if you could throw it away fast enough—faster than seven miles per second—it might escape the pull of the earth and never come back. They refer to that as "escape velocity."

Whenever our kids release their helium-filled balloons, they stay right where they are. But not on planet Earth! Theirs go *away from* the planet. However, dried-up leaves, when disconnecting from their trees, consistently travel toward the earth. Balloons and leaves travel in opposite directions—balloons upward and leaves downward. And they have "clouds" that seem to be exempt from either rising or falling. It's all so spooky!

Earthlings put their hands between the earth and certain objects to move those objects away from the earth. They call this "lifting the thing up." And, then, when certain things are released, they will travel back toward the earth. Sometimes, inside the boxes they call "houses," these things will hit the floor and occasionally roll. People complain about the ways in which these things roll to the farthest places, and under what they call the "lowest" pieces of furniture. All of this follows a mystical law that they call "Murphy's."

They plant crops. While they are growing, most of these plants expand outward, upward, or away from the earth. However, some reclusive things, which they call "peanuts" and "potatoes," decide to stay hidden in the earth, refusing to come out into the open. People must go after them in a special way that they call "digging." It's all so inconsistent and strange!

They have a game called "volleyball" that we cannot play, because, for us, when the ball goes over the net, it does not come back and offer us a chance to hit it again. After all, there's no such thing as "up" and "down" where we live. But their balls curve in mid-air and come back toward their planet where players can hit them again. I tell you, Tok, I'm glad to be back in our country. Long live the ZGE!

For Discussion

1. Lacking the experience of weight while in orbit, how can astronauts in spaceships "weigh" themselves to see if they are gaining or losing weight?

2. In a ZGE, how would one drink water out of a glass? Take a bath? Lean against a table? Fry an egg in cooking oil?

CHAPTER 12

On the Meaning of Meaning

"From the beginning the universe is a psychic as well as a physical reality."
—*Thomas Berry*

"I am interested in meaning because it is the essential feature of consciousness, because meaning is being as far as the mind is concerned."
—*Physicist David Bohm*

As a boy, I lived on a farm where my father kept and marketed honeybees. I would sometimes approach the rear of the hives after dark and quietly place my ear against them. Within those boxes I could hear thousands of bees busily at work, fanning the moisture out of their day's collection of nectar so as to turn it into honey. They seemed never to really rest. Indeed, even in the darkness of night, sensational wonders were going on in those hives while most people went about their business oblivious to them.

I learned that, during the lifetime of a worker bee, it gathers about four teaspoonfuls of nectar, which eventually becomes just one teaspoonful of honey. The total distance flown by bees in producing a pint of honey is approximately three times the distance around the earth! I relate these facts as an introduction to the subject of meaning, for I can never look at bees or taste honey today without experiencing a deep inner glow—one of familiarity and gratitude. After all, my family's livelihood once depended on countless numbers of those honeybees. Thus, I fully believe that *the meaning of bees and honey* is different for me than it is for most others.

Just what is it that we are trying to convey when asserting that something is meaningful, or that it has high meaning? Although vital to everyone, meaning is not a concept that can be easily defined. Dictionaries seem to explain it in terms of significance, only to define significance in terms of meaning. When exploring the meaning of meaning, then, we quickly come to the realization that meaning itself is a pretty slippery concept. Still, meaning seems to be the topmost concern of everyone everywhere, for whoever makes choices of any kind does so on the basis of meaning. On what, then, is meaning based?

I recently asked several university colleagues of mine, all of whom were highly experienced in philosophical thought, "What is it that one is implying when declaring that something is meaningful?" The following are three responses given to me by theology professors:

I think of meaning in terms of hope. In my way of thinking, things are meaningful if they are in some way life-promoting.

Whatever I judge to be fulfilling in the here and now, this is what I call meaningful.

Things and events are understood as meaningful to whatever extent they have practical relationship, or association, with other things in our system of values.

But is meaning only relative to our lives, or can it be an absolute? Could trees have meaning if there were no people around to know trees? Or, considered from another point of view, do meanings depend *only* on the outlook of the beholder? Perhaps we should take into account both the relative and the absolute for a proper understanding of meaning.

As to the relational aspect of meaning, let us consider the following example: What educated people normally call a "pencil," if given to a Papuan tribesman, would likely be viewed as a kind of blowgun dart. He might instinctively attempt to fit the pencil into the bore of his blowgun, imagining its point piercing the feathers and hides of birds and animals. The concept of pencil as an instrument for writing would probably not enter his mind until he had witnessed someone using the object for the purpose of writing.

We can reasonably say, then, that specific things have different meanings to different people depending on their backgrounds. Furthermore, we can look at a given object as having different meanings for an individual person under different circumstances. During a fire, for example, a piano that is ordinarily understood as a massive musical instrument might suddenly be viewed in a practical new way as a device that enables one to climb up within reach of a high window to escape the burning building. Certain people are especially talented in reading different meanings into specific things. They include the ingenious ones who improvise in such ways as converting empty oil drums into stoves, or making pajamas out of discarded sugar sacks.

Thus, meaning can be viewed as very relative, depending not only on particular places and circumstances but also on the backgrounds and hopes of the observers. Author-psychiatrist Viktor Frankl, a survivor of the Nazi death camps, has written:

> A man's search for meaning is a primary force in his life and not a secondary realization of instinctual drives. This meaning is unique and specific in that it must and can be fulfilled by him alone. Only then does it achieve a significance which will satisfy his own will to meaning.[1]

It is indeed true that the concepts of significance and meaning are very closely related. There are very strong elements of subjectivity in Frankl's idea of meaning, and rightly so, for the perception that things, situations and events take on new meaning involves our recognition of them in different ways. Perception of meaning always depends on our backgrounds, talents, and visualizations, on where we are coming from and where we hope to go in search of fulfillment!

One of the many ways in which people perceive meaning is by correlation. Nature is full of similarities and contrasts that help us to understand things in terms of other things. We often speak of a thing being similar to something already known. Things perceived as opposites impress us greatly. We also notice things exhibiting numerous gradations between opposites, such as various shades of gray between black and white. Thus, it is not surprising to find philosophies such as Taoism[2] that stress harmony within the forces of nature by recognition of antitheses.

According to Taoists, nothing in life is recognizable without its corresponding opposite. Opposites are said to complement and give meaning to each other, and a Taoist would say that light has no meaning without darkness, nor does bigness have meaning without smallness, nor male without female, nor rain without dryness. Right is meaningless without left. High is meaningless without low. Good has no meaning without evil. And, ultimately, one cannot appreciate life without awareness of death. An eternal pattern of opposites giving meaning to each other swirls throughout the Taoist view of meaning.

However, a new question arises: Is meaning altogether relative? Hardly so, because while grasping for the roots of meaning, we will always be drawn toward an absolute. Instinctively, we seek a kind of ground or "ultimate fundamental base,"[3] a foundation on which all worth is established, including not only the worth of meaning, but also the worth of the concept of worth. In this context, we must stress the importance of the human psyche, the recognizer of worth. The psyche has risen from the cosmos, which itself is our primary referent or ground of believability.

The cosmos is the immediately experienced initiator of truth recognition. By concentrating deeply on the things of this world with humble abandonment to truth, we come to recognize meanings on finer levels and treasure them as refreshing and life-promoting. Even were all our worldly possessions to be lost, the haunting search for meaning would still be our principal concern.

In the physical sciences we attempt, as much as possible, to focus on what we call "objective physical reality." But our attempts to express the behavior of nature are necessarily subjective acts. Thus, personal experience cannot be dismissed in quests for meaning even as we strive to express physical laws in ways that are free from human prejudice. We can never totally put aside our subjectivity to become wholly absorbed in objective otherness, for we are enmeshed in a kind of web within which we struggle in subjective ways to be objective.[4]

Yet, we continually strive to fly in spirit, transcending beyond the limitations of our self-centeredness, with the cosmos serving as the well-grounded base for that engagement. The cosmos becomes the substance, situation, and opportunity that we utilize when rising above and going beyond what we call the ordinary.

When trying to understand the meaning of meaning, then, we must take such factors into consideration. The cosmos itself impels us to imaginatively fly out in spirit from its bosom of believability. We do so with full assurance that we are, indeed, moving toward genuineness if we are persons of good will. Our innermost concern for genuineness in things and events is also a zest for an absolute, this absolute being the inescapable factor in our every consideration of meaning.

Inasmuch as our concern with meaning is by nature cosmos-based and endlessly linked to all of life, we might well think of the cosmos as a kind of triggering reality. That is, it invites us onward in a never-ending earthly search for meaning. Thomas Berry frequently refers to the universe as "our primary referent," for it invites, initiates, induces, and inspires our quests for meaning. We, in turn, exhibit our faith that fulfillment is there to be found. In our odyssey, we pursue meaning while knowing that the chase itself has meaning. Always we search for the wonder of goodness, including the goodness of success in our most trivial acts, such as in lifting a spoon to stir our coffee.

It is recorded that world-renowned physicists Albert Einstein and Albert Michelson met for the first time in the year 1931. Einstein asked Michelson why had he gone to such great pains in measuring the speed of light. Michelson simply replied, "Because it was fun!" He, like all of us, craved the satisfaction of discovering more about nature by way of what he considered to be elegant entertainment.

Whenever we search for meaning in a particular situation or thing, invariably our concern is whether it enhances our life. Relatedness is an essential element here—relatedness among mundane things such as a baseball, a bat, and a glove, with bigger things in mind. Negatives, too, are included in this consideration, such as our vigilance toward things that ought to be avoided.

When searching for meaning, we customarily do so in the context of personal experiences—our own as well as those of others that we imaginatively assimilate and, in effect, turn into our own. Assimilated experiences in today's world include walking on the moon, sky-diving, watching faraway events on television, engaging in computer-animated games, becoming lost in the past (such as by watching *Gone With the Wind*), and countless other such engagements. We who look at the world imaginatively in the context of both personal and assimilated experience sooner or later come to ask: Toward what does experience converge? In the totality of life's activities, what is the ultimate outgrowth of "world experienced"? (This subject will be addressed in the last chapter).

Our search for meaning is analogous in some ways to putting together a jigsaw puzzle. An individual piece does not achieve its full significance until it is seen to fit snugly against its neighboring pieces. Then all of the pieces viewed together reveal a scene that makes sense on a higher level. The deeper we look, the more profound the meanings discerned.

Search for the Absolute, in my opinion, is at the root of every assessment and affirmation we make. To the extent that we perceive the deity as Goodness, as Truth, or as Beauty in its fullness, we also perceive that God is in some way at issue in every evaluation. This connection is evident to deeply spiritual people who sense that universals are discernible in particulars. Awareness of this is harbored in the subconscious and it enables us, while experiencing the world, to extrapolate our intuitions through the endlessly expansive realm of implications. Consistently, such a process leads toward the infinite. What glorious experiences are to be found along the way!

Not as easy to see, however, is the fact that suffering and discomfort, too, harbor transcendent meaning in our awareness of "world." In the Christian consciousness, Christ is understood as God-become-human. In him rests the fullness of meaning. Once God identified with the makeup of the world, then everything everywhere assumed a new dignity. All of creation became related in some more profound way with God through Christ.

Christians believe that God is everywhere. But since the coming of Christ, the term "everywhere" is understood in a new and profoundly different way. The coming of Christ has shifted the meaning of everything, and because of this nothing anywhere or at any time can be properly viewed as unrelated to Christ.

Christians visualize themselves as constituents of the cosmos who strive to live in the likeness of Christ. They aspire to embrace nature holistically. This means including even the weak, the sick, and the poor in their worldview of what is meaningful in life. They are aware that, no matter what they do, the highest meanings of their actions relate to Christ, and that there is no way to evade this fact.

To a civilized person, what we call a pencil is only a pencil...until the Papuan tribesman shifts its meaning by declaring it to be also a blowgun dart. When he puts it to use, it then *becomes* a blowgun dart—or, perhaps, a blowgun dart that can also be used as a pencil.

A pebble is only a pebble until we recognize that it is worth our time to pick it up, to examine it as a particular entity, and then to view it as an object "worth throwing." To throw a pebble and watch where it lands is to dignify both the pebble and the environment where the pebble finally lands. Any attention we give to so-called "insignificant" things reveals our deep-seated belief that they are worth our time. They take on meaning for us even when we engage them in a trivial or casual manner.

Because of who we are and what we value, the plastic cups on our kitchen shelves are no longer mere plastic. When we recognize their rigidity, their lightness of weight, their shapes and sizes, their thermal insulating properties in terms of suitability, this recognition elevates them in our world of meanings. If, additionally, we understand the chemistry of their substance, then we see them with greater insight than do others and are able to crown them additionally with our recognition of their invisible but nevertheless very real molecular structures. We wash them, display them, and hope to reuse them. By doing so, we declare them to be worthwhile and valuable for the future. We subconsciously affirm them as having significance above that of the things we discard. However, lest we forget, even the things we discard are meaningful entities which, having served their time, we lay to rest...hopefully with dignity, such as by recycling them.

The cosmos is a colossal entity in a state of waiting, with all things, including ourselves, biding their time toward fulfillment, toward the "coming next step" in their existence. An object of leather, rubber, air, and thread lies around uselessly until a boy recognizes it as a soccer ball. Then it "comes alive" (in a poetic manner of speaking), winning his attention, stimulating his imagination toward having fun. As far as humans are concerned, the things of the cosmos are "merely there" until our recognitions/meanings are bestowed on them.

The elevation of things into the world of meaning effects a remarkable transition. And it is we ourselves who do the elevating. This is often expressed in a trivial manner by simple commands. A mere directive such as "Give me that set of keys" affirms the metal-become-keys and the keys as enablers of security and therefore as worth receiving. The directive also implicitly affirms the giver of the keys and the manufacturers of the keys as people who respond to the needs of others. These hidden people become associated in meaning with the things we sense as worth giving, worth receiving, and also, alas, as worth ignoring. In the minds of people willing to visualize, the implications of things are always far-reaching. Thinking of them in meaningful ways confers recognition on them.

Returning once again to the Christian awareness of this world, a new and paradoxical ethic arises for the human family, one that is rooted in the basic human practice of recognizing meanings. Simply put, the surest way toward highest transcendence is to serve others in a new cosmic order wherein the humble have been lifted into the realms of topmost meaning. In the Christian view of the world, the Creator came into the human family among the poor. His body was of material substance, just like ours. Thus, material everywhere has been elevated in dignity. In the Christian awareness, God through Christ has become a relative of everything everywhere, and now has something in common not only with people, but also with things. In its own way, everything of this world, in one way or another, has a relationship to Christ who is the Way.

Notes

1. Viktor Frankl, *Man's Search for Meaning: From Death Camp to Existentialism*, Beacon Press (1959), 99.

2. Taoist texts date back to about 600 BCE in China. Taoism teaches simplicity of life and conformity with the cosmic order.

3. Karl Rahner, S.J., *Foundations of Christian Faith*, trans. William V. Dych, S.J., Seabury Press (1978), 87. Rahner writes of God as "the transcendental ground of the world [who] from the outset embedded himself in this world as its self-communicating ground."

4. An example of this is the following: In the case of the periodic table of the chemical elements, we could say, objectively speaking, "Atoms exist in such a way that their chemical properties are repetitive as progression takes place from the lighter elements toward the heavier elements." The element of subjectivity, however, becomes apparent when we say more openly: "*By our looking at atoms* from the viewpoint of their behavior, *we have come to understand that* they exist in such a way that their chemical properties are repetitive as progression takes place from the lighter elements toward the heavier elements."

For Discussion

1. Give an example of something that is commonly discarded but that you think has beauty.

2. Can the meaning of things change without the things themselves undergoing change?

3. Do you think that things are wondrous in themselves? Or is it that they become wondrous when humans like ourselves recognize them as wondrous?

CHAPTER 13

In Search of Absolutes

Earthquakes are frightening experiences, even for people who are outdoors and far removed from buildings that might collapse. To us who normally think of the earth as the mainstay of our stability, the jolting and swaying we experience during a quake shakes our very foundations, both literally and figuratively.

Our trust in the so-called stable earth runs deep, so deep that our most profound religious beliefs are sometimes expressed in words descriptive of the earth's stability. Examples of this trust abound in the Old Testament, as in the psalms:

> "You set the earth on its foundations,
>   so that it shall never be shaken." (Psalm 104:5)

Such acclamations seem to reveal a human need to trust in an entity that is final and absolute. This can also be noted in our methods of speaking, for we habitually cite things and events said to be the first or the last of their kind. Record-setting events are another case in point. We speak of the most, the biggest, the highest, and we employ adjectives such as *world-class* or *space-age* to communicate what we deem as respectable and worthy of respect and acclaim. The exclamation: "We're number one!" comes to mind.

The point is that we like to use superlatives to absolutize or crown wondrous things and events. It is a tendency that runs deep in our hearts, revealing within us a consistent desire for standards that are self-sufficient. To exclaim "Tell me the truth!" or to assert "Of course I am right!" is to imply the existence of an absolute. That is, we sense that we are in touch with an ultimate referent of a sort whenever we are correct, and at least to some degree out of touch with it whenever we err. Our alignment with this absolute seems to be of crucial importance if we are to know ourselves as genuine.

Seemingly, too, our closeness with the absolute lies at the root of our belief that we have basic worth and dignity. To the extent that we speak the truth, we do so as people in touch with what is real, laudable, proper, genuine, and necessary. Our understanding of *who we are* versus *who we ought to be* includes consideration of these ultimates.

Furthermore, we reveal our implicit belief in a moral order whenever we use terms like *ought, right, truth, integrity, justice,* and *trust.* The sincere use of words such as these, words that defy physical description, reveals an underlying response to transcendent mystery. In fact, whenever we perform a willful act of any kind, we do so in recognition of an absolute, for to act in a responsible manner is to behave with awareness that the outcomes of our acts are viable. Our acts live on. They are recognizable as having inherent worth. They somehow *count,* and we are indelibly linked to them in the sense of being answerable for them. Indeed, the mere fact that our acts linger in our memory seems to confirm that they count and can influence even our future.

I clearly recall how, as an undergraduate physics student, I reacted to the statement that all motion is relative. My instructor hastened to explain that speedometers on automobiles are designed to record our speeds relative to the earth. But the earth itself is not at rest; it is moving relative to the sun. And the sun is moving relative to our galaxy, which in turn is moving relative to the universe. As far as we can tell, there is no fixed reference point in the whole of the cosmos. Therefore, my instructor explained, there is no such thing as an absolutely fixed frame of reference anywhere!

Upon hearing this, I felt slightly disenchanted. After all, in answer to the vitally important question "Where am I?" I could only say that I am aboard a planet orbiting a star located in the "great somewhere." I can say little else concerning my overall whereabouts, for no one knows where the cosmos itself is located—if, indeed, the term "location" makes any sense when applied to the cosmos as a whole.

I first became aware of a kind of relativity when I was a young boy living in the country. My father was a railroad agent at the time. Occasionally, he stood on the tracks near dusk, lit a sheet of paper, and swung it back and forth to flag down an approaching train. The steam locomotive engineer would respond by blowing his whistle twice. Invariably, we saw the two puffs of steam squirting from the whistle five to ten seconds before we heard its sounds. We understood that our experience of hearing the whistle differed in time from that of the engineer. When the sounds of the whistle were a present experience for us, they were clearly a past experience for the engineer.

As it happens, a relativity based on sound differs somewhat from one based on light. The oddity about light is that its measured speed is

ALBERT EINSTEIN
(1879–1955)

Considered among the foremost geniuses of all time, Albert Einstein was born in Ulm, Germany, and had early problems in school that once led to his dismissal. After slow beginnings, he finally earned a Ph.D. and became a citizen of Switzerland. He was hired as a "technical expert third class" patent examiner in the Swiss Patent Office in Bern, where he found time to compose several scientific articles, three of which started new branches of physics. The one that immediately shook the world of physics was his treatise entitled "On the Electrodynamics of Moving Bodies," published in 1905. In it, he departed from the accepted laws of Newtonian physics and proposed a world where space and time are relative and the speed of light is absolute.

As a Jew who criticized Hitler, Einstein was quick to sense that he had no future in the regions dominated by the Nazis. He and his wife, while visiting America, decided to remain there when Hitler came to power in 1933. Learning that the Nazis were beginning to conduct research on development of an atomic bomb, he wrote an urgent letter to President Roosevelt in 1939, alerting him to that fact and suggesting that the United States begin research on nuclear fission without delay. Yet, he was a pacifist and later intensely advocated world peace through nuclear disarmament.

In America he spent the remainder of his career in a research position at the Institute for Advanced Study in Princeton, N.J. Einstein strongly believed that if the human race was to continue, it had to create a moral order.

invariant. That is, it travels through a vacuum at 186,000 miles per second regardless of the motion of the observer. It makes no difference whether the observer, while measuring the speed of light, is traveling against the movement of the light or with the movement of the light. Either way, the speed of light turns out to be the same, this odd behavior proving to be a great absolute in nature as a whole. The discovery of this fact by Michelson and Morley in 1887 startled the world of physics.

The behavior of light is of great concern to us because it is the basis for so very many of our observations of nature. Whereas most scientists were bewildered by Michelson and Morley's revelation, Albert Einstein simply accepted the fact that nature behaved in that way. With impressive docility he accepted cosmic behavior exactly as observed, and then proceeded to formulate the subtle implications of that behavior. Out of Einstein's leap of faith came his theories of relativity, which gave us vastly deeper insights into the fundamental behavior of physical reality.

As a result of Einstein's contributions, our previous concepts of space and time were greatly modified. The old ideas of absolute space and absolute time were replaced by the new concept of variable space-time in a universe where the speed of light is the absolute. Stated briefly, his theory of relativity said that basic quantities such as mass, length, and time, as measured by persons in different frames of reference, can differ radically depending on the relative velocities of the observers.

Most of Einstein's predictions based on relativity have now been verified. Yet certain diehards maintain that his perceptions are glaringly incorrect. They seem to forget that his view of relativity as a whole was based from its outset on observed invariance in the speed of light, a phenomenon that contradicts common sense (if by "common sense" we mean a type of understanding based only on the slow-moving things of everyday experience). By advancing beyond the common ways of thinking and placing his faith in what light tells us about itself, Einstein brought forth an inspiring new understanding of the physical universe. In relation to the cosmos, he did what mystics sometimes do in the realm of spirituality when they abandon themselves to the Spirit, emerging with new and profound intuitions. But, alas, people can always be found who belittle mystics.

Following Einstein's publication on special relativity in 1905, certain extremists began insisting that *everything* in the world is relative and, for that reason, people cannot be truly certain about anything. Often this was, and still is, done in jest. For example, I have overheard students who are studying Einstein's relativity for the first time exclaim in a spirit of fun, "So, then, everything is relative! Nothing is absolute! I cannot be certain of anything, including even myself! What's the use of trying?"

Clearly, if Einstein believed that we can know nothing with certainty, he would never have published. His view of relativity was that we can, if we take the proper factors into account, observe nature and draw valid conclusions that reliably correspond to reality. Einstein's theories of relativity serve to unify or bring into congruence nature as observed from different frames of reference, particularly where super-high speeds are involved.

As was explained earlier, we never see the stars or anything else as they are "right now" in our precisely local present moment. We on earth see the sun as it was eight minutes ago, and the other stars as they were eons ago, more or less deep in antiquity according to their distances from us who observe them in what we subjectively call the present moment. An ontological question arises at this point: Can we validly think of the universe in terms of all of its components existing in an all-pervading present moment, in an "absolute now" of a sort that is independent of time understood relativistically? To put the question in other words, can we say, "Simultaneous events are occurring everywhere in the cosmos *right now*"?

We certainly act as if we believe in an absolute present moment that pervades the whole of the cosmos.[1] We act this way whenever we utter a thought such as "I wonder what is happening on the star Alpha Centauri *at this moment!*"

Einstein did not deny the validity of our thinking of the present moment existing as an absolute entity throughout the cosmos. He implied, rather, that our physical observations of the universe are subject to a unique kind of time-space error, which, if we are not careful, can lead us into false interpretations of physical events. We can liken this to a fisherman who stands on a pier intending to spear a fish. As the light travels from fish to man, it undergoes bending when passing from water to air, thus giving the man an erroneous idea of where the fish is located. If, however, the man knows how refraction behaves, he can then aim his spear at some point below where the fish appears to be, thus successfully spearing the fish. By knowing his physics and compensating for the bending of the light, he can indeed overcome a deceptive appearance.

All in all, there is no necessary contradiction between the spiritual belief in an absolute and the scientific preoccupation with the relative, which, in its own way, addresses the absolute. Indeed, science addresses the absolute in numerous subtle ways. As stated by the late physicist David Bohm:

> The essential characteristic of scientific research is . . . that it moves toward the absolute by studying the relative in its inexhaustible multiplicity and diversity.

By necessity, we utilize the cosmos when describing our understanding of how the cosmos behaves, for no one can leave the cosmos and describe nature as if viewing it from the outside. Thus, we have no choice but to express what we understand of nature's behavior from within the context that nature allows.

But the cosmos behaves in consistent ways that are so very reliable that we fully place our trust in them as if responding to an absolute. Indeed, we repeatedly stake our lives on the reliability of nature, on its behaving today as it behaved yesterday, last year, and a thousand years earlier.

Recognition of nature's consistent and mysterious ways is the very footing on which physical scientists base their work in seeking to describe nature's behavior with accuracy. Possessing a yearning for truth at least in this regard, the physical scientist addresses the cosmos in the overall framework of spirit. Although working with "mere material," as some would call it, the scientist is yet a spirit-driven individual who is engaged in uncovering and revealing what is true. Always and forever, the search for truth throughout the cosmos—even though we must perform it locally—is a colossal engagement with the Absolute.

Note

1. Physicists are now speculating about the existence of "superluminal connections," wherein an entity of some kind manifesting speeds greater than that of light is thought to keep things "in touch" with one another.

For Discussion

1. What do you understand the word "absolute" to mean?

2. Is it possible for God to be other than absolute?

3. What do you think scientists mean when they speak of "absolute zero" in the world of temperatures?

COLLIDING GALAXIES!

Do galaxies ever collide? Indeed they do, as shown in the above photograph taken by NASA's Hubble Advanced Camera. Located three hundred million light-years away, the pair is known as NGC 4676 and nicknamed "The Mice" because of their tails. Computer models suggest that we are seeing these two galaxies some 160 million years following their closest encounter. The lengthy arm at the right is curved, but appears straight as we view it edge-on. It is believed that the pair will eventually merge into a single giant galaxy.

CHAPTER 14

Is Christ a Cosmic Absolute?

"In him all the fullness of God was pleased to dwell,
and through him God was pleased to reconcile to himself all things."
—Colossians 1:19–20

Webster defines *absolute* as "free or independent of anything extraneous; complete in itself." It is a traditional belief of Christians that Christ is the absolute Savior of the world: that he and he alone can effect total salvation or victory over death, and that his salvation is eternal and irrevocable. Christians also believe that Christ, as the absolute Savior, remains with us as a living presence in the world even though he died centuries ago. Following are but a few of the numerous scriptural passages referring to Christ's absolute quality:

"Very truly, I tell you, before Abraham was, I am." (John 8:58)

"Father, glorify me in your own presence with the glory that I had in your presence before the world existed." (John 17:5)

"All things came into being through him, and without him not one thing came into being." (John 1:3)

"No one comes to the Father except through me." (John 14:6)

The basic Christian leap of faith holds that God became a historical human person, a man known as Jesus of Nazareth, who walked the earth like we do. According to this belief, God assumed a human nature like our own, with matter united to spirit in him as it is in ourselves. God did not masquerade under a human form. Rather, God *became* a human person in Jesus. Like us, Jesus experienced hunger, fatigue, sexuality, elation, temptation, disappointment, wonder, and everything else we ourselves ordinarily experience in our own circumstances.

When viewing the Incarnation from an evolutionary point of view, one might understand that living entities on earth evolved to the point of becoming rational and self-reflective. Then, at some "acceptable time," God lovingly chose to appear on earth as one of them. In Christ, the Creator became personally identified with the substance of the world: matter, energy, space, and time. God became human, stated theologian Karl Rahner, so that the cosmos in all its countless forms could be properly recognized as "the environment of God's own materiality."[1] Humans, in particular, could then be viewed as God's own kind in ways extending beyond the Old Testament notion of "God's people." Our understanding of concepts such as these can open the door to reverence for all things and all people everywhere.

Rahner viewed the Incarnation as "the necessary and permanent beginning of the divinization of the world as a whole."[2] He insisted that Christ, the absolute Savior, is "the moving power of the movement toward the goal,"[3] and that goal is Christ himself who is God. In Jesus, God became one of us so as to provide a clear example of how we ought to live. Through Jesus, God spoke in audible words and visibly acted with a human body like ours. Jesus' earthly existence made it possible for us to think and speak of the deity in easily understood worldly terms.

Theologians tell us that the message of Jesus and his person are inseparable. Paradoxically, they add up to a single consideration, rather than two, for while Jesus spoke of resurrection and life, he identified himself as *the* resurrection and *the* life. He spoke the truth and insisted that he himself *is* the truth. This implies that whenever he spoke, he was communicating his personhood to others. *His claim implies that whenever we speak the truth about anything, we are in some way communicating him to others.* And, furthermore, whenever we observe the endless individual and collective truths of things everywhere around us, we are in some way that is fundamental, essential, and immediate also in touch with Christ in whom both their individual truths and their relational truths rest. It is precisely here that our odyssey of searching for meaning in the Christian awareness of "world" is rooted. Here, also, is where the term "Christness" as it relates to the world comes forward in our awareness.[4]

Jesus spoke of himself and his followers as one—the vine and the branches. He also made it clear that in his Father's sight humanity now includes God and ourselves together as one in himself. In the Roman Catholic liturgy we find this prayer:

"Father, . . . so great was your love that you gave us your Son as our redeemer. You sent him as one like ourselves, though free from sin, that you might see and love in us what you see and love in Christ."[5]

Admittedly, prophets and saints of many religions have appeared on the earth. In the fully Christian view, these people are worthy of imitation precisely to the extent that they express Christ in this world, for Christians see Jesus as the foundation of Wonder, the one in whom all meaning in this world rests. Such has been the central mystery of Christianity, setting it apart from religions that deny the world, or view it as base, evil, or devoid of meaning. Because of the Incarnation, Christianity has emerged as a religion that is supremely transcendent, yet also firmly down-to-earth in the Lord Jesus.

Because of God having personally identified with this world through Christ, Christians cannot properly think of their encounters, either with one another or with material things, as purely secular. Neither can Christians properly believe that material alone *is* God, or that God *is* the sum total of material at large (pantheism). But Christians can, and with good reason, speak of material as dignified, uplifted, and made holy by God's intrinsic unity with the material world through the physical body of Christ. And as to the scientists who address matter so analytically, Christians can correctly speak of them as focusing deeper-than-usual attention on the environment of God's own materiality. Indeed, the Incarnation has made it possible for us to achieve higher levels of appreciation for everything everywhere. And so, through Christian eyes, it is in relation to the Incarnation that the very definition of what a scientist is attains its highest meaning.

Existence comprises the foundation of all that is knowable, for there is nothing more fundamental than existence. God, who is Existence in the Fullest, is embedded in this world as its communicating Ground. This means that the world, including ourselves, rests in God and can be fully understood only in terms of God—in terms of its truth, its presence, its value, its goodness, its loveliness, its wonder, its thrust, its outreach, its goals, and other such transcendent qualities. Thus, in a fully theistic awareness, it is the recognition of God in everything everywhere that gives the concept of "world" its majestic and foremost meanings.

The Christian perspective of this world clearly promotes this point of view. Christians view Christ as God-become-world, as God committed to, or embedded in, the cosmic situation so as to draw us fully away from what is merely situational and toward what is absolute. In the words of Rahner:

"It is Christ who signifies the beginning of the absolute self-communication of God which is moving toward its goal."[6]

In view of this, Christians have very good reasons to speak of the cosmos—in particular the material world—in exalted ways. For God is irrevocably and personally involved in matter-energy and space-time through Christ. In becoming one of us, God assumed the constituents of this world in the same way that we possess them by way of our bodily makeup. Thus, with Christ in mind, we can speak of the world in terms of God, and speak of God in terms of the world—views that are wholly compatible in the Christian psyche. Anything, anywhere, at any time can occasion the bringing of Christ to mind.

Christians who are well versed in the New Testament eventually come to recognize the truth revelations of the cosmos and those of Christ as somehow intertwined. They come to see the revelations of the cosmos that are written in science books as resting within the kingdom of Truth. Assuming that Jesus is who he said he is, namely the Truth, then the truth of anything, anywhere, at any moment relates directly to him.

Maturing Christians sense something of the Lord abiding in the makeup of the world that they observe, utilize, talk about, and take for granted. Often their concern for the betterment of the environment is linked to this perception. In their eyes, Christ is the fulfillment of a wondrous creation, which, because of his intrinsic involvement in it, is endowed with an everlasting dignity. Christ is a cosmic Absolute in whom all things and all situations rest for all time. As St. Paul wrote:

> "He himself is before all things, and in him all things hold together." (Colossians 1:17)

In other words, everything everywhere is related to Christ. And, because of this, everything everywhere is precious.

Christ is a cosmic absolute. If he and the Father of creation are one, how could it be that he would not be a cosmic absolute?

Notes

1. Karl Rahner, S.J., *Foundations of Christian Faith*, trans. William V. Dych, S.J., Seabury Press (1978), 197.

2. Ibid., 181.

3. Ibid., 195.

4. Further considerations of "Christness" appear in chapter 33 on "The World as Holy Sacrament."

5. Preface, Sundays in Ordinary Time, VII.

6. Rahner, *Foundations of Christian Faith*, 193.

For Discussion

1. What are your thoughts with regard to viewing the universe as "the environment of God's own materiality"?

2. Is there anything in this world that one can truthfully say is unrelated to God?

3. What, then, do you understand the word "secular" to mean?

CHAPTER 15

On the Shepherdhood of Matter

"Blessed be you, mortal matter,
you who one day will undergo the process of dissolution within us
and will thereby take us forcibly into the very heart of that which exists."
—Pierre Teilhard de Chardin

Dimly recognized but almost never mentioned, there is a quiet shepherdhood at work in the substance of this world, in matter and in energy everywhere and at every moment. It is an active shepherdhood, embedded and ongoing in all things and situations as they offer us opportunities and possibilities simply by being what they are. Even space itself, considered as the receptacle of objects, partakes of this quality. And all of this comes under the governance of time, which also shepherds us relentlessly into new beginnings.

Our ability to both understand things and imagine what they might do for us are essential human qualities that lie deep within us. We think: How can I utilize this or that? What opportunity does it offer? What options does it hold open for us? How ought I to treat it if I want it to be there later on for me to call upon it again? Questions such as these illustrate our ongoing awareness of potential in things as we view them in terms of their readiness to energize us.

The quantities and qualities of things, their suitability in enabling us to extend ourselves beyond our personal limitations, these all linger in our awareness. Matter routinely invites us to call upon it to liberate us. The sooner we admit this, the sooner we realize that we are under the spell of a shepherdhood of a sort.

Our awareness of this shepherdhood begins in early childhood as we test matter in countless ways. At first we familiarize ourselves with the basic behavior of various substances—such as wood, metal, glass, rubber, cloth, cardboard, plastic—which we encounter in the form of such things as toys, floors, spoons, books, napkins, sand, marbles, and mud pies. We repeatedly test the level of tolerance of things by pushing them, as safely as possible, to their limits. We come to understand their inertia when bumping into them or when feeling them bump into us. We learn that the more massive they are, the more difficult it is to stop them or get them moving. The softer they are, the more suitable they are as surfaces on which we can sleep. The steeper they are, the better they might be for sliding downhill in a sled, provided they are not *too* steep. Clearly, we become attuned to the ways of matter.

Matter and its twin, energy, display ways of their own on which we come to depend every moment of our lives. Some people use things and simply enjoy them. Others use them, enjoy them, and philosophize over them (as we are doing at the present moment), identifying deeply with their essences and becoming entranced with their appeal. We may envision the countless fascinating ways in which matter stands before us as a great and endless orchestration, a conglomeration and interweaving of shapes, sizes, densities, colors, strengths, flexibilities, malleabilities, viscosities, conductivities, reflectances, and other such characteristics. And, if we are scientists, we engage in a further kind of contemplation. We revel in matter's crystalline structures, in its magnetic susceptibilities, its ratios of charge-to-mass, its secondary electron-emission capabilities, its radioactivity, its exotic smoothness, and very much more. There is something in matter-energy and space-time for everyone.

People appreciate the dependability and submissiveness of matter, the ways in which we can rely on it to do, as we say, "what it is supposed to do." We commonly assume matter to be quietly present to us as if awaiting our next move. Soil, for example, is understood by farmers as awaiting cultivation so that it may serve them as material that will nurture the growth of food. Sometimes also, if we are sensitive enough, matter can be understood as imploring us to leave it alone. A well-constructed violin needs no further work, and labor performed on such an instrument would likely be of disadvantage. "If it ain't broke," we sometimes joke, "don't fix it!"

The task of mentally establishing things in the psychic order as manifestations of newfound meaning and truth-expression is a truly magnificent engagement. And one would be mistaken to think of this activity as unique to educated adults, for the very young and the uneducated also engage in wondrous creativity, even if in casual and unsophisticated ways. Uneducated Peruvian Indians making sandals from discarded automobile tires exemplify this, as do schoolboys making paper airplanes and launching them from upstairs windows.

As mentioned in a previous chapter, our innermost awareness of who we are is linked to what we do with what is available in our surroundings. For example, we cannot mail a letter without simultaneously knowing ourselves as persons who sometimes mail letters. Thus, things and situations contribute greatly to defining us as individuals. People often say who they are in terms of what they do with material things: a syrup engineer, a water inspector, a concrete man, a truck driver, a steeplejack. It is worth noting that, while recognizing our work with things, we should not dismiss our engagements with them as acts that are purely secular. For in the consciousness of those who recognize their time on earth as limited and precious, no engagement with things can be properly conceived as lacking in personal meaning.

Things enter deeply into our spiritual world as we identify with them. Within our hearts we harbor a hope/expectation whenever we use a screwdriver, reach for a telephone, open a newspaper, or eat an apple. In using things, we respond to opportunities we recognize in them. We recognize the suitability of their substance, their form, or their location for rescuing us from certain deprivations of ours.

Good expectations are a kind of subliminal prayer of hope. When we have expectations of things, we are banking on their reliabilities according to what we discern as their capabilities. Even when we choose *not* to utilize a certain thing, such as a particular bridge that spans a river, we are nevertheless comforted by the options that it holds open to us, perhaps for use at a later time. The options that unused things offer us enrich us. They widen our psychic world, enabling us to understand that we have alternatives, such as choosing this rather than that, more rather than less, later rather than now.

So fundamental are the favors offered to us by matter that they are taken for granted and almost never discussed. But they are firmly and vitally implanted in our consciousness and, therefore, ought to be explicitly affirmed, at least now and then.

We do in fact subliminally endow matter with some degree of vitality when we say that particular things do or do not "live up to" our expectations. Such statements reveal that we habitually apply notions of expectation/evaluation, recognizing in them a sort of passive kindness, an advocacy that can enhance our lives.

Within the innermost realms of our psyche we customarily rank things in terms of their perceived importance. Mentally, we endow them with meanings in the context of their availability and reliability, in terms of the promise they offer us. A police officer does this when purchasing a bulletproof vest, and he reaffirms it every time he "calls upon it" by strapping it on.

Occasionally, we may even address things aloud while utilizing them. Children, especially, are disposed to express their expectations out loud, such as when they exclaim, "OK, gloves! Keep my hands warm!" or "All right, batteries, don't go dead on me now when I need you most!"

These childlike verbalizations may strike us as ridiculous, but they are *explicit* expressions of our everyday *implicit* hopes/expectations of things.

Numerous things immediately display to us what we can expect of them: a plow for plowing, a basketball for bouncing, a book for reading. When matching these things with our particular needs, we place our trust in what we see as their capabilities. We remember these capabilities. They remain within our inner world of value judgment, elevated or demoted according to countless evaluations we bestow on them. These evaluations range from "vital and life-promoting" to "damaging or deadly."

Clearly, our relationship with matter is a two-way street. We not only draw upon the usefulness of things, but we also are drawn out of ourselves in spirit by them. This happens as we address all sorts of things everywhere with our needs in mind:

"Sustain me," we utter from the depths of our hearts to the air, the food, and the drink that we ingest.

"Come to my aid," we say deep inside ourselves to our computer keyboard, can opener, lawn mower, microwave oven, automobile rear-view mirror, and endless other things.

"Give us another chance" is our unspoken message of hope directed toward our grindstones and our spare tires as we call upon them to sharpen our tools for new beginnings or to enable us to continue on our journeys. Tools, by definition, rescue us from our tool-less predicaments.

We routinely direct optimistic pleas and expectations toward automobile service stations, furniture repair shops, and electric razors. In these cases we welcome their abilities to resupply, refurbish, or refresh. On the other hand, there are times when we address with guarded hopes and expectations the things from which we wish to remain at safe distances: perhaps a snake lurking near our campsite, or a fallen electrical power line. We may not express any of these sentiments audibly, yet our hopes/expectations are definitely there, lying at the root of our relationships with many things.

All these things, in turn, silently communicate back to us our dependence on them. The full extent of their message is apparent only when we are in cadence with them, particularly to the extent that we are willing to enfold them into our psychic world of appreciation for what they offer us. Otherwise, their higher implications go unheard as we make use of them in thoughtless ways, deaf to their meanings and messages.

The mere presence of things bestows meaning on our senses. Were it not for the presence of things to be seen, our eyes would have no meaning, and those who presently enjoy good vision would be no better off than those who are blind. Vibrating matter, the source of sound, gives meaning to our ears. Were it not for the motions of matter, we would, in effect, be totally deaf. If there were no motions, there could be no sound. If there were no sounds, then there could be no music either.

To utilize things in appreciative ways, then, is to be in touch with a divine-like quality of shepherdhood presented to everyone in the context of time. Also, when using things in harmony with their ways, their truth and their life-promoting wonders, we gain understanding of ourselves as peaceful inhabitants of the cosmos.

Maturing Christians come to relate the shepherdhood of matter with that of Christ. It is a new interpretation of "world," one in which we sense that his shepherdhood is still at work today, not only in the spirit of his followers, but also in the very substance and situations of creation which they address at every moment. They seek to recognize this quality of the Messiah in everything everywhere, particularly in the ways that various things can offer us distinctive kinds of liberation.

Because of God's intimate connection to creation through Christ, everything everywhere is endowed with transcendent meaning for believers. Beginning at the moment of the Incarnation, the meaning of all creation has shifted. Connections with Christ, and therefore with God, are discernible in all things. In the Christian mentality, the ways and the truths of things flow back to the idea that God engaged all of this world through Christ. Thus, when we address this world with a good heart, we are proceeding both spiritually and materially along a path taken by God through Jesus.

What, then, does it mean to recognize a thing as suitable, of good use, or "there when we need it"? The Christian answers: It is to see God's goodness resting in that thing. Christians who sense the charisma of creatures also perceive Christ's call to life by way of the good that creatures hold open to us.

But what about situations where matter and time seem to work so firmly against us? How can God's shepherdhood be recognized in situations of physical adversity? Admittedly, we encounter all sorts of obstacles, and they seem at times to stifle our very freedom. In response to the above questions, we might well consider the following reflection from Pierre Teilhard de Chardin's "Hymn to Matter":

> Blessed be you, harsh matter, barren soil, stubborn rock: you who yield only to violence, you who force us to work if we would eat.
> Blessed be you, perilous matter, violent sea, untamable passion: you who, unless we fetter you, will devour us.
> Blessed be you, mighty matter, irresistible march of evolution, reality ever new-born: you who, by constantly shattering our

mental categories, force us to go ever further in our pursuit of truth.

Blessed be you, universal matter, immeasurable time, boundless aether, triple abyss of stars and atoms and generations: you who, by overflowing and dissolving our narrow standards or measurements, reveal to us the dimensions of God.

Blessed be you, impenetrable matter: you who, interposed between our minds and the world of essences, cause us to pierce through the seamless veil of phenomena.

Blessed be you, mortal matter: you who one day will undergo the process of dissolution within us and will thereby take us forcibly into the very heart of that which exists.

Without you, without your onslaughts, without your uprootings of us we should remain all our lives inert, stagnant, ignorant both of ourselves and of God. You who batter us and then dress our wounds, you who resist us and yield to us, you who wreck and build, you who shackle and liberate, the sap of our souls, the hand of God, the flesh of Christ: it is you, matter, that I bless.[1]

Teilhard's graciousness toward matter exemplifies an excellent attitude for us to take, regardless of the particular form of matter. Consider transistors, for example. These tiny devices are the outcome of certain carefully selected materials having been treated with delicacy and special care. We find transistor production personnel wearing masks, white coats, and cotton gloves, all out of consideration for what the elements germanium and silicon can do. The required respect we exhibit toward such substances, in expectation of what they can do for us, frequently and appropriately borders on reverence.

When we speak of matter's shepherdhood, inventions and innovations readily come to mind. Inventions are, after all, the outcome of inspirations. Through them, newfound expressions of being enter our world, each manifesting yet another mode of God's presence among us. Inventions exhibit the art of applied knowledge, actualized from an endless set of possibilities that have awaited their time. As such, they can enable us to recognize the life-promoting presence of God in ways that are immediate, practical, and particular.

In the Christian understanding, creative activities spring from the Holy Spirit. This process can be visualized in terms of God's Spirit using creation to originate good ideas in the minds of receptive individuals who visualize new possibilities among things.[2]

Artists and inventors should balance creativity and discretion in utilizing available materials for the human good. In this way society may become the beneficiary of a proper surrender to the lure of things within the context of grace. The result—transcendent delight in good inventions—is part of what it means to be fully alive. For, as long as the concepts of technology and responsibility are linked, inventions become God-centered expressions of our humanity in tune with the cosmic mystique.

On a grand scale, the rigors of inventing and engineering exemplify a kind of cerebral evolution. It is this evolution about which Robert O'Connell writes:

> ...We must remember to rejuvenate our vision of what is going on by jettisoning the hard and fast "natural versus artificial" distinction. [If we] think of the so-called "artificial" as the "humanized natural" [then] the airplane becomes the genuine analogue of the bird; the submarine, the analogue of the fish; and so on down the list of human inventions. But, by the same token, we are jolted suddenly into seeing roads, air and sea lanes, radio and telegraph lines as constituting a "veritable nervous system for humanity" as a single living organism, the substrate of a "common consciousness of the human multitude" looked on as a whole.[3]

In considering matter's shepherdhood, we must also acknowledge its vulnerability, for, paradoxically, the substances and situations of a shepherding world exhibit astounding submissiveness. Trees stand in readiness to be sawed into lumber and take care of our needs for shelter. Gravel and concrete submit to our trampling, coal to our mining and burning, mountains to our digging of tunnels through them. Bread submits to the knife. Matter offers us opportunities to mash it, melt it, pulverize it, decompose it, bend it, blend it, twist it, vaporize it, grind and recycle it many times over. Poetically speaking at least, by being what it is matter is thus open to a kind of death and resurrection at our hands. It is through the inherent submissiveness of atoms that scientists can now use particle accelerators to smash them and know some of their innermost secrets.

Christians grow in their understanding that God has granted matter the dignity of servitude, the chance of participating, at a level proper to itself, in the scriptural teaching that those who serve others are of high importance in the kingdom of God. The greatness of matter, then, resides in its innate readiness to serve. It continually comes to our aid with specific supports being offered by specific materials, each in its own recognizable way.

Paradoxically, then, a Christian may rightly envision the world as an outcome of a self-imposed constraint on the part of the Creator. In this view, God, in creating the world, has subjected God's own self to particular commitments from which escape is impossible. The existence of creatures indelibly affects both the physical and the metaphysical worlds, because once things exist, it becomes thenceforth impossible for them not to have existed. Even God could not make it otherwise without violating the Divine Nature, which is to ceaselessly sustain without revocation all that has been created.[4]

This vulnerability of God by way of creation is also apparent in other ways—by exposure to criticism, for example. Once the cosmos spiritualizes to the point of containing rational creatures such as humans, it thereafter becomes an arena of evaluation/assessment. Humans can praise God for creation. But humans can also challenge God, such as by questioning or mindlessly discounting the transcendent qualities of this world. They can ask why the world was not created better—free from human suffering, for example. This is a dilemma analogous to what we ourselves experience when, regardless of what we do, we cannot please everyone.

The evaluating minds of humans repeatedly test creation in endless ways. A well-balanced ax, a reliable windshield wiper, properly matched loudspeakers, a suitable pair of binoculars, a sufficiently insulated house—these exemplify qualities to which we are drawn when pursuing excellence in the entity we call "world."

Christians who prayerfully search for the highest meanings, who see Christ as God's direct involvement in the stuff of creation, can discern something of Christ in the submissive devotion of matter. Indeed, something of the vulnerable Shepherd is present in the vulnerabilities of matter as it "goes about its business" of doing so much for so many despite the many times it is shamefully treated, such as during the infamous burning of oil wells in Kuwait in the year 1991.

Scientific discoveries and good technological achievements, then, represent occasions for celebration. After all, joy over discovery is a mode of proclaiming that what was previously absent to human consciousness has at last become present. Such joy is expressive of a proper materialism, one that is centered in God. It renders thanks to God by way of grateful attention to the individual things of creation, much as a child might delight in a doll or a bicycle.

All in all, the ways in which we understand our past encounters with specific things allow us to chart our future according to the opportunities they offer. We address glass, trusting that its refraction will save us from blurry vision. We lift a cup to our lips expecting that it will contain our coffee and cream in ways far better than our hands alone could. We turn to the materials of sphygmomanometers in hopes of obtaining accurate readings of blood pressure. How wonderful it is to be able to do these things!

The world of television offers us a special case. Television enables us to see faraway places through the eyes of others, such as the distantly lo-cated camera operators. And, even more, television reaches into worlds extending far beyond this example.

News commentators stand in front of television cameras, gazing and smiling at the camera lenses as if they were actually speaking to the glass itself. Everyone in the studio accepts that behavior as both proper and commendable. I suspect that people would look at me strangely were they to see me smiling and talking to a doorknob, or to a bottle of glue. But smiling and talking to a curved hunk of glass in the barrel of a box called a camera would be considered totally acceptable!

The experience of speaking to material substances while having faith that they are highly developed conveyors of human personhood is certainly a psychological breakthrough in human history. Imagination plays a significant role in that transcendence.

A truly amazing aspect of television lies in the everyday fact that viewers do personally relate with what they see on the screens before them. I am especially referring to the ways in which TV link-ups are conducted between two persons in cities far apart. Each "sees" the other while talking back and forth, each addressing fluorescent-screen repre-sentations of the other. In a unique sort of way, both are using materials while transcending materials—the phosphors of the picture tube screens—as if personally engaging each other in the substances of the screens. What marvelous substances we have at our disposal today. How heavily we have come to lean upon their quiet and mysterious invitations to come forward through them into the light of life!

Notes

1. Pierre Teilhard de Chardin, S.J., *Hymn of the Universe*, Harper and Row, Inc. (1961), 68.

2. It is said that the television pioneer Philo Farnsworth, as a teenager in high school, conceived the idea of "sending pictures by radio" while culti-vating a field. He afterwards gazed at the rows he had plowed and asked himself if it might be possible to photograph a scene and send pictures of it by radio, one line at a time (like the furrows), and reassemble them else-where in order to reconstruct the scene.

3. Robert J. O'Connell, S.J., *Teilhard's Vision of the Past*, Fordham Univer-sity Press (1982), 60.

4. Thomas Aquinas referred to this as follows: "It is obvious, moreover, that God cannot make the past not to have been, for this, too, would entail a contradiction; it is equally as necessary for a thing to be while it is, as to have been while it was." *Summa Contra Gentiles*, Bk. II, Ch. 25 (15).

For Discussion

1.   Distinguish between proper and improper materialism.

2.   Do you agree that inventions are special gifts, the outcomes of inspirations or "good ideas" given to individuals by God?

3.   Would you describe yourself as a person who, when using such things as light bulbs, scissors, and sidewalks, has an ongoing awareness of the people who made these things?

CHAPTER 16

Divine-like Attributes of the World

"God truly waits for us in things, unless indeed he advances to meet us."
—*Pierre Teilhard de Chardin*

Matter is truly a great mystery. There is so much of it around us that it we can easily think of it as merely ordinary. Considering, however, that most of the space in the universe is empty, there are reasons to think of matter as the exception rather than the rule in the cosmos as a whole.

Indeed, it is truly wondrous that anything at all exists. And here we are, humans, the cerebralized part of the world, navigating the surface of the sphere called Earth while responding to our cosmic situation at every turn. But we easily miss the deeper meanings of the world unless we become discerners, dreamers of a sort who reflect deeply on and "tune in" to the implications and innuendoes of things.

Relating to the Creator through creation is a highly spiritual engagement. It is a noble pursuit, not to be confused with false materialism or pantheism. Those who reverence God by way of creation endow matter with the dignity it deserves as an outward expression of the inner life of God, as an immediate and tangible sign of the Lord's presence in the world. Our spiritual relationship with matter, energy, space, and time is, at its best, like that of a young lady who kisses her engagement ring, not because she adores the substance of the ring, but because the ring brings to mind the love of her husband-to-be.

In relation to this, the movie entitled *The Sound of Music* comes to mind. At the film's opening, actress Julie Andrews dances atop a hill while singing "The hills are alive with the sounds of music!" Indeed, they seem very much alive, for Andrews herself makes them come alive in the consciousness of the viewers. She is bubbling over in her awareness of the beauty of the mountains, with their colorful flowers and cool fresh air, and she communicates that spirit to her audience with great immediacy.

Many Old Testament passages convey the same sort of vibrancy. For example, the scriptural writers speak of the Lord as "having pitched a tent for the sun" in the heavens, as if to confirm that the sun is occupying its "rightful place" in the skies. Other passages depict the mountains as "leaping like rams," and trees as "clapping their hands." One passage goes on to exclaim: "See, the Lord GOD comes with might, and his arm rules for him; ... who has weighed the mountains in scales and the hills in a balance!" (Isaiah 40:10, 12).

There is much poetic beauty in expressions such as these. "Here comes with power the Lord!" seems to proclaim both grandeur and excitement. It expresses not only the strength of the Creator, but also the awe of the beholder, and it does so in a deeply poetic manner. Also, there is something fascinating about the Lord "weighing" the hills and the mountains as if on balances. As a physicist I am tempted to ask myself, "What kind of balance was being used? Was God working with gravitational masses or with inertial masses?" I smile a little as I realize that I am missing the point, for the writer was simply expressing joy and wonder in a figurative manner. A literal understanding was probably the last thing in mind. Yet I enjoy playing around with literal meanings just for fun!

As a teacher, I enjoyed listening to the sentiments of my physics students, especially toward the end of the course. One semester I asked them to write down their feelings about the object of their studies—the cosmos. The following are several of their responses:

> "The cosmos is amazingly beautiful and awesome! It seems overpowering and everlasting. I feel extremely privileged to see it all."

> "I am amazed at how everything seems to be related with everything else, and how they all interact with one another. I feel that everything we do has an effect on every other thing in the universe."

> "I have mixed feelings about the cosmos, as if I am rich in having it, yet poor and subjected to being swallowed up by it."

But of all the student responses received, the following was my favorite:

> "I am *pleased* with the cosmos. I cannot come up with anything else to take its place."

In return, I described to them some of the personal insights I have gained by listening with discernment to the inner and outer workings of the cosmos across the course of fifty or so years. Some of these I would like to share here, subjective though they may be:

First and foremost, the things of the world communicate to us that they are our ongoing companions. The buildings, the roads, the trees, the grains in wood, the street signs, the flagpoles, the sidewalks, and the doorknobs—they invite our attention and help liberate us from desolation and inner privations just by being there. Hour after hour, year after year, they are with us when we fall asleep, and we fully expect them to be present when we awaken. They seem to say: "Moment after moment, century after century, we perdure! And it is to your advantage that we do!"

Interestingly, whoever studies scripture can relate to a similar understanding about God. For while we understand God as immediately present in the here and now, the words "here" and "now" have no meaning without considerations of the world's makeup. Thus, we might well view the steadfast, abiding presence of things as a *God-like quality* that they possess.

An ancient utterance made by the Lord reveals this abiding presence of God in terms of all that exists. In the Book of Exodus is recorded that God said to Moses:

"I AM WHO I AM." (Exodus 3:14)

The wonderful outcome of that revelation was that Jews, Christians, and countless others have since grown to understand that the concepts of God and existence go together as one. Thus, by reason of the universe existing, it thereby possesses a special aliveness—not just in the sense of some of its creatures being biologically alive, but also in the sense that all of it is *ablaze with the expression of Being.*

Over time, people of religious faith grow toward understanding what this means. They see that, in knowing particular things in this world, they are engaging the Being of the Lord expressed in particular ways through the being of particular creatures. They may recognize something of God in any creature anywhere at any time. Indeed, the Being of God and the charisma or attractiveness of creatures work together closely in our ongoing awareness of the world.

Scientists are particularly entranced by nature's overall changelessness, within which all the changeables operate. They typically formulate and articulate the so-called "physical laws" from their observations of nature's consistent behaviors. The laws of nature say in what manner the changeables in nature behave according to seemingly fixed rules. A falling ball, for example, undergoes change in its location and speed, but it does so in a very predictable way. Science would make no sense if things behaved differently under similar conditions. Therefore, the description of nature's recurring behavior is the concern of the scientist, even to the point where scientists will articulate the orderliness of disorder. They speak in terms of entropy, chaos theory, and statistical mechanics . . . loads of fun for those who love higher mathematics.

But of highest importance in our experience of "world" are the psychological factors. In particular, we must understand that, if things were to behave differently at different times, we would quickly lose our ability to cope with reality. We can imagine what a terrifying experience it would be to jump upward if we were uncertain about gravity pulling us back to the earth today just as it did yesterday. Our basic faith in nature, then, is very much one of acquired confidence in the fact that nature will never change "the rules of the game."

In behaving today as it did yesterday, and tomorrow as it did today, matter whispers a kind of promise to which each of us has become attuned since infancy—the promise not to betray. Like the God of Israel, the cosmos has, in effect, entered into a covenant-like relationship with us. It is an age-old pledge of dependability not to forget the needs of its people, who themselves have arisen out of the trustworthy behavior of its substance and circumstances.

For reasons such as this, we can view the physical sciences as providing us with noble descriptions of dependability from of old. I understand the scientist's infatuation and preoccupation with the stuff of the world as an affirmative response to fascinating qualities of reliable behavior implanted in the makeup of this world. Because the cosmos is the medium through which we learn to expect consistency of behavior, I shall cite *dependability from of old* as another God-like quality of the cosmos.

Awareness of physical laws can engender within us feelings of contact with the expansive and the everlasting. The very manner in which material substance exhibits dimension, the manner in which it participates in the apparently endless progression of time—these draw us beyond the present. And, to the extent that we flow with the implications of all this, we are drawn onward and outward from our local involvements into more spiritual dimensions.

Let us imagine that we stroll to the edge of a cliff. The way that we come to a halt and steadfastly refuse to take another forward step says, in effect, that we have heard a message from that particular arrangement of matter. Another example is that of a stop sign at a highway intersection. We recognize the sign, its shape, its color, and its letters, and we discern what it is telling us. We then respond to it exactly as if it had voiced a verbal command.

This, then, leads us to recognize another God-like quality of the cosmos; namely, that it *evokes humility*. It demands and obtains responses of respect from us. In certain instances, it compels us to back off from our previous engagements, to change our plans for our own good. It does so in much the same way as the Lord demands surrender, sacrifice, and restraint from us in the life of our spirit. The cosmos induces us to be cautious, to exercise moderation while navigating in freedom. So, too, does the Lord demand surrender from us while offering us freedom.

A further God-like quality of the cosmos lies in the fact that it is *expressive of a way* in the natural order of things. The cosmos whispers a message: "I am the way. You must necessarily relate to me *on my terms, not yours!* You must design your ships, your planes, and your radios in harmony with my behavior if they are to float and fly and tune in stations. You can never take a break from considering my ways without risking disaster." It is a message to which not only engineers, but also artists, truck drivers, and everyone else must be attuned. Carpenters, for example, are attuned to it when, after listening to the ways of wood, they avoid driving nails too close to the edges lest the wood split. All who are careful in ways such as this are listening at least subconsciously to nature telling them about its ways.

It is both interesting and wonderful that we hear similar messages (but on higher levels of awareness) from scripture. God's biblical revelations may come to us more directly than those of nature, but God's spoken themes are much the same as the implicit ones we have just considered. Take, for instance, these words of Christ:

I am the way, and the truth, and the life. (John 14:6)

Taken together, all these messages imply that nature and God cannot be bypassed, or, perhaps, that nature is modeled after the Creator. And we know that we must come to terms both with God and with nature, or else we will suffer disaster. "I am the way" is an ongoing message that comes to us from both the physical and the metaphysical, the realms that constitute the totality of our world experience.

The makeup of the world helps us to understand truth as active from the beginning, reaching forward through all things and for all time. Truth is what we expect to be in touch with whenever we know anything, anywhere, and at any time.

By being attentive to the things of creation in a spirit of appreciation, we become attuned to their truth-communications. The chairs in our rooms seem to say, "I truly am designed for you to sit in." The water we drink seems to whisper, "I truly am able to satisfy your thirst and help sustain your life." Even persons who regard themselves as atheists respond with faith to these non-verbalized "I truly am" messages. Endlessly throughout each day, we respond to messages from every creature around us, each of which by its very existence affirms non-verbally "I truly am" in its own particular way.

There is a growing emphasis in today's theology on the fact that response to the world is ultimately response to God who is embedded in a creation that is representative of, and yet distinguishable from, the Creator. In the Christian psyche, this awareness is a furtherance of Jesus' teaching that the ways in which we treat each other are the ways in which we treat him. Our proper responses to the gift of "world" are in fact responses to God by way of responses to God's outer expressions.

Christians who grow in spirit gradually come to sense that all human action is fundamentally a response to the "I am" uttered by God to Moses, and to the "I am the way" uttered by Jesus to his followers. They respond to the world as persons addressing the Way and the Truth of God through the ways and the truths in things, situations, and people. Thus, I include *the ways and truths expressed in all things* as yet another God-like quality of the cosmos.

Another important way of viewing the constituents of the cosmos is to see them, like G. K. Chesterton did, as coming to our rescue by eliminating deprivations. This view might best be understood if we start with man-made things, which generally are outgrowths of our perceived necessities. Clocks rescue us from our native inability to accurately keep track of time. Shovels and hoes save us from the difficulty of cultivating our gardens using only our hands. Musical composers look to French horns for rescue, for deliverance from their inability to evoke French-horn moods without the use of French horns.

And so it is with glass and plastics. Simply by being themselves—transparent to light and rigid while at room temperature—they exhibit refraction properties that enable us to enjoy corrected vision. Whether or not they are actually used, they nevertheless remain available to rescue us from defective eyesight. Similar words could also be spoken about coat hangers, shoes, buckets, paper clips, tape recorders, milk cartons, and everything else that we normally use. Each comes to our rescue in its own specific way, exhibiting a messiah-like quality proper to itself. Our use of them is our silent acknowledgment of this.

Even unattractive things open up to us a kind of natural salvation. Our vision, even if it were 20/20, would be devoid of meaning unless we also had things, including the unattractive, at which to gaze, for there is no advantage to having sight unless we also have things to see. So, also, with our other external senses; they would have no meaning and serve no purpose unless what they sense does in fact exist. But they do so charmingly exist. Again, a time to celebrate!

Nature, like God, provides us with countless second chances. Throughout our lives we stub our toes, burn our hands, break our bones, and habitually look forward to a recovery that generally does take place. From our earliest life experiences we learn that nature is repeatedly forgiving—perhaps seventy times seven times, the number mentioned by the Lord. Even time itself, by its passing, helps us to forget devastating experiences. Nature seems to tell us—through the healing power of living matter, the coagulation of blood, the knitting of bones—that everything will be all right once again. Yet we dare not become presumptuous of nature's healings, for nature, like the Lord, can also manifest an awesome non-healing finality on those who are heedless of warnings. Prudence is required.

Undoubtedly, the healings we experience through nature help us to visualize the merciful forgiveness of God. For, within limits, nature repeatedly gives us new beginnings, just as God does. When we are in need of nature's healings, we instinctively turn to splints, bandages, medicines, x-rays, wheelchairs, and the passing of time itself. Consequently, another God-like quality of the world is the *healing forgiveness* that it frequently offers us.

To Christian eyes the entire movement of the cosmos as God-expression is recognized as resting in the Savior. Christ is the Ground of meaning reaching back into the meanings of the world even in pre-Christian times.[1]

Christians recognize humility as a characteristic of God, who became vulnerable to cosmic conditions in becoming one of us. Christ in his humanity submitted to worldly conditions right up to the moment of his death. But, amazingly, whatever reveals God also conceals God. This can be the case at times with our human sufferings. Through our sufferings we sometime become blind to the presence of God, even though they are a way for us to relate to God all the more.

In Old Testament days, suffering was understood as a curse. With the coming of Christ, however, suffering has taken on new meaning. By accepting our sufferings, we can more deeply identify with Jesus, who personally experienced rejection and pain. Because of him, our physical and emotional wounds have more than mere secular significance. By way of him who is the Way in its totality, distinctions between what is secular and what is sacred have long ago faded. With Christ in mind, we have reasons for viewing the concept of "secular" as a negation that fails to stand on its own.

Note

1. As explained by Karl Rahner, S.J., in *Foundations of Christian Faith*, trans. William V. Dych, S.J., Seabury Press (1978), 194: "Insofar as a historical movement lives by virtue of its end even in its beginnings, because the

real essence of its dynamism is the desire for the goal, it is completely legitimate to understand the whole movement of God's self-communication to the human race as borne by this savior [Christ] even when it [the movement] is taking place temporally prior to the event of its irrevocable coming-to-be in the savior."

For Discussion

1. Would you agree that constancy (or consistency) of behavior in the ways of nature is a necessary condition for our mental health?

2. Is it possible for God not to be interested in this world or, perhaps, to have a "low level" of interest in certain segments of this world?

3. Is there a limit to the number of truth statements that you can make about yourself?

CHAPTER 17

Teilhard's Divine Milieu

"Quite specifically, it is Christ whom we make or whom we undo in all things."
—*Pierre Teilhard de Chardin*

Among the outstanding modern thinkers in the area of creation spirituality is Pierre Teilhard de Chardin. Teilhard, who died in 1955, was a French Jesuit paleontologist who labored for years in China. He was deeply involved with both the spiritual and the material realms of creation, and he wrote:

> Throughout my life...the world has gradually taken on light and fire for me until it has come to envelop me in one mass of luminosity glowing from within.[1]

Two of Teilhard's best-known books are *The Phenomenon of Man* and *The Divine Milieu*. The former contains his philosophy for reconciling Christian theology with scientific evolution. In that work he refers to Newton's universal gravitation as merely the shadow of that which really moves nature, namely spiritual attraction. He identifies love in all of its subtleties as "the direct trace marked on the elements by the psychal convergence of the universe upon itself."

Teilhard viewed the cosmos as an ensemble, as a grand performance of an organized interdependent whole. He described the tendency for matter to unite beginning inside molecules and extending outward through such forces as gravitation, evolution, sexual attraction, parental instinct, and compassion. He believed in continuous creation, the process wherein "God makes things make themselves." Just as science views matter and energy as two aspects of the same cosmic entity, Teilhard visualized matter and spirit not as separate or side-by-side, but rather as a single cosmic entity exhibiting two faces, which he called "matter-spirit." He visualized the earth as having evolved through three basic phases or movements through what he called *spheres* or *envelopes of containment.*

The first sphere, called the *lithosphere,* includes the earth's crust. The second he called the *biosphere,* or "envelope of life." Then comes the *noosphere,* or "psychic envelope," which includes the mind. Thus, Teilhard envisioned the evolutionary passage from pre-life to life as movement through a critical threshold, followed by movement through a second critical threshold when rational humans appeared as "cerebralization" of the earth. Humans, he insisted, are the components through which the cosmos reflects upon itself in self-awareness. With the appearance of rational creatures, the world was said to have been "hominized." In such a state, evil became a possibility because rational beings could choose to turn away from the natural upward swing implanted in the ways of the cosmos. Therefore, the appearance of humans meant that the sacred process of "cosmification" could now be not only enhanced, but also obstructed.

In *The Divine Milieu* Teilhard visualized the cosmos as containing divinity. He believed that the stuff of the world offers us endless glimpses of the Creator here on earth and summons us to response. The call comes to us in countless ways: through our likes and dislikes, our desire to earn a living by involvement with things and people in different places and times, our restlessness to invent and innovate, to sketch, to devise, to repair, to concoct, to bypass, to overcome, to endure, to reach out, to try our hand at various things. Through it all, Teilhard surmised, a little something of God is immediately attainable in our work and our art.

Teilhard also implied that there is a certain basic dishonesty in people who, while utilizing the material world, speak of it with disdain. He insisted that each of us "makes our own soul" by what we do with this world on a daily basis. He reminded us that, by virtue of creation and Incarnation (God becoming one of us in time), nothing here below is profane for those who know how to see. By responding, then, to the divine as perceived in things, circumstances, and people, we become the living prolongation of God's creative power in a world where creation is still occurring. Thus, whereas most of modern Christian thought had advocated living with eyes fixed not on this world but on eternity, Teilhard had a different approach. He emphasized matter in all of its forms and circumstances as revealing God's modes of presence in this world.[2] He insisted that atoms and molecules of every conceivable kind manifest the grandeur of the Creator everywhere.

Teilhard's writings shocked some of his religious peers. Initially the Vatican's *L'Osservatore Romano* objected to his writings about "a cosmic nature" in Christ. In later years, however, the Vatican seemed implicitly

PIERRE TEILHARD DE CHARDIN, S.J.
1881–1955

A French Jesuit biologist and paleontologist, Pierre Teilhard de Chardin labored for many years in China, where he collaborated in research that resulted in the discovery of the Peking man (1929). Among his best-known works is *The Phenomenon of Man*, in which he elaborates an extensive system of thought that addresses science and evolution in the light of Christian teachings. Teilhard believed that the cosmos undergoes increasing "complexification" expressive of God's "worldification" through Christ. Another book of his, entitled *The Divine Milieu*, explains our closeness to God through every thing and every place at every moment.

In his anthropocentric and somewhat mystical thought, humanity is regarded as the key to the universe. In Teilhard's view, this will ultimately lead to an "Omega point" in which the world, including all of humanity, will be taken up in God through Christ.

to accept his cosmic views of Christ, although serious misunderstandings of his writings still persist in some religious circles. Similarly, some of Teilhard's fellow scientists objected to his writings as "unscientific nonsense tricked out in a variety of metaphysical conceits." Today, however, many scientists of religious faith acclaim his writings as illuminating the interconnectedness of science, philosophy, and religion. Over time, Teilhard's writings have found an increasing acceptance among scientific and religious thinkers, establishing his status as a seminal thinker.

The writings of Pierre Teilhard de Chardin are unusually all-inclusive and holistic, and thus not always easily understood. Books have been published to help readers understand what he was saying. Above all, he dared to replace the view that the things of the world are "merely secular." Instead, he asserted that this world is sacred, that God is alive, ongoing, and recognizable in every segment of creation. And he insisted on a Christ-centered perspective in harmony with the Christian gospels. As he wrote:

> Do away with the historical reality of Christ, and the divine omnipresence which inspires us becomes, like all dreams of metaphysics, uncertain, vague, abstract, without tangible authority over our thinking, without moral imperatives for our lives.[3]

Considerations such as these are of great relevance to anyone wishing to view the world from truly broad perspectives. Especially attractive are Teilhard's ways of unifying science and religion, so I will repeatedly draw on his thinking as we broach other topics in this volume.

Notes

1. This quote, as well as this chapter's opening quote, are found in Teilhard's *The Divine Milieu*, Harper and Row (1960), pp. 13 and 123, respectively.

2. One factor that must be taken into account, though, is the way in which the term *world* has been used (and misused) with many different meanings across the centuries in religious and philosophical writings.

3. *The Divine Milieu*, p. 112.

For Discussion

1. Do you believe that your responses to the world will turn out in the end to have been responses to God by way of the world?

CHAPTER 18

Nature's Lengthy Threads of Simplicity

*"A theory is the more impressive the greater the simplicity of its premises,
the more different kinds of things it relates,
and the more extended its area of applicability."*
—Albert Einstein

Prior to Newton (1642–1727) it was assumed that the laws governing earthly things were different from those governing heavenly objects. During Newton's early years it was typical of philosophers to reason in the following way: Every object in the universe has a proper place, a location that is specifically its own. It strives to remain in its place. Things near the earth are terrestrial. When lifted and released, they fall to the earth because "down" is their natural place. Objects such as the moon and stars, on the other hand, are "of the heavens." They remain in the skies because "up" is their proper place.

Newton showed that the so-called "terrestrial laws" and the "heavenly laws" were in fact different manifestations of the same behavior, two faces of a single all-inclusive law. He recognized the pull on the falling apple, on the orbiting moon, on people (giving them their weight), and on everything else in the world as a single behavioral pattern at work among all things. This unifying insight by Newton—his universal law of gravitation—marked a tremendous leap forward in the understanding of simplicity throughout nature. However, it had yet to be verified. Would the characteristic of "universality" be found to hold true for things farther away from the earth than our moon? Would it be found that the same behaviors of nature at work locally were also at work very far away? And, finally, could it be shown that the modes of behavior of things in this world are the same today as they were in antiquity?

Astronomers later observed that the behavior of gravity near our earth applies universally, throughout all of space. Photographs of stars made years apart clearly reveal how binary stars (star pairs) spin around

105

each other under the influence of their gravitational attractions—like two small children holding hands and running around each other.

Powerful new insights also arose in the late nineteenth century thanks to a new technology known as spectroscopy. Spectroscopes showed that light emitted by different kinds of atoms displayed different characteristic colors (spectra). It was as if particular kinds of atoms had their own distinctive fingerprints.

At the same time, astronomers observed that light from the sun revealed strange color patterns, as if coming from an unknown gas. They questioned whether certain atoms unknown on earth might be present in the sun, or whether atoms in the sun were behaving differently from the same kinds of atoms on earth. One gas in particular came into question. Its spectral display, coming to earth from the sun, was unlike any emitted from known things on earth. Given the name *helium*, this gas had atoms of a kind that were later discovered on earth as a by-product of radioactivity. Later still, helium was found in abundance in certain wells in northern Texas.[1] Through observations such as this, scientists extended the principle of universality to include the emission of light.

The spectral patterns of light from the stars became especially important. Spectral comparisons showed that the distinct colors emitted by hydrogen in the most distant stars identically matched those from light emitted by glowing hydrogen in our laboratories. Thus scientists realized that nature's fundamental behavior everywhere was very likely the same. The characteristic of universality had held true when comparing the local behavior of light with its behavior at great distances and over vast spans of time. For, when observing light from the very distant stars, we are observing the behavior of atoms that emitted that light five to ten billion years ago. Radiation from atoms in stars, although weakening in intensity with distance, does not decay in its characteristic properties when traveling over enormous distances and across eons of time. That is, the recognizability of the radiation from a given kind of atom remains intact regardless of the *where or when* of its source...more evidence of simplicity throughout nature.

In regard to this quality of universality throughout nature, its behaving in consistent ways always and everywhere, Nobel laureate physicist Richard Feynman remarked in a poetic vein:

> Nature uses only the longest threads to weave her patterns, so each small piece of her fabric reveals the organization of the entire tapestry.[2]

Driven by a kind of instinct or taste for simplicity, today's physicists are attempting to interrelate the whole of the universe. They search for a single set of laws to describe the behavior of everything—from elementary particles to galaxies. Obsessed with this so-called Grand Unification Theory (sometimes called GUT), cosmologists such as England's Stephen Hawking seek to discover a single great principle that underlies the whole of physical reality.

As envisioned, the Grand Unification Theory would be a universal, all-encompassing principle that would account for the physical behavior of everything everywhere throughout the course of all time. If scientists could thoroughly describe the behavior of a single thing in terms of that universal principle, they would have essentially related it with every other object, situation, and event in the whole of the world, past and present. By applying a single universal principle to individual situations, they would consistently end up with the same results that we presently obtain by utilizing numerous, apparently separate laws formulated in the various branches of physics. If we ever accomplish this, we would realize interconnectedness at its fullest.

Interestingly, numerous scientists agree that a single universal principle would have to be simple rather than complex. Perhaps it would be analogous in some way to Shakespeare's "to be or not to be," concise, yet all-encompassing. Feynman, in the passage cited above, went on to share what he called "an unscientific view" concerning the cosmos. He claimed that the reason it is possible to guess from one part of nature's behavior what the rest of nature is going to do is that "nature has a simplicity and, therefore, a great beauty."

An interesting shift in the history of science occurred in 1924, when French physicist Louis de Broglie suggested the existence of *matter waves*, which he imagined as associated with moving things. His thoughts on this can be paraphrased as follows: Nature loves harmony and symmetry. Matter and energy are two faces of a single entity in the makeup of this world. If radiation energy (in motion) displays wave-like behavior, then matter in motion should also display wave-like behavior. If photons of light have wavelength, so also should moving matter, such as electrons (and even baseballs) in motion have wavelength.

Reportedly, de Broglie's mentors at the University of Paris were hesitant to award him a doctoral degree. Three years later, however, physicists Clinton J. Davisson and Lester H. Germer showed that his speculations were well founded. Experimenting with streams of electrons, they reflected them off nickel crystals and were surprised to see them bouncing off in specific directions, exhibiting wavelengths in agreement with the predictions of de Broglie. Davisson and Germer were able to account for this behavior in terms of the moving electrons instigating matter waves.

Connections between beauty, simplicity, and truth were being made in many other areas of thought, including the world of psychology. For example, two or three years after de Broglie proposed his hypothesis, Carl Jung wrote this: "When a thing suggests beauty or harmony in its form, it always has more to do with the truth than when it is ugly." Jung was specifically referring to symmetry in a pattern he had sketched relating to dream analysis, but his message was similar to that of contemporary physicists and other scientists.

Consistently across the centuries, those who have sought to give us insights into the question of who God is have referred to God in terms of simplicity. Thomas Aquinas (1225–1274) wrote:

> God is wholly single and simple in himself. The human mind, however, which does not see him as he is, cannot know him except through many concepts.[3]

It is one of the strange facts of life that in order to study the smallest—and presumably the simplest—of particles, we must erect the largest of scientific instruments. To wit, the sub-atomic realm is presently turning into a jungle or "zoo" of two hundred or more sub-elementary particles, yet atomic physicists need to work with gigantic atom smashers of various types in exploring the world of super-tiny quarks and leptons.

So numerous are the varieties of elementary particles being discovered that we are classifying them in categories of "charm," "beauty" and "flavor"—jargon once regarded as proper only to mystics pursuing simplicity in the world of the spirit. Suspicions abound that the Grand Unification Theory might be near, and the term "finding God in the atom" is gaining usage even in science. Although exciting, this does not mean that those who utilize such terminology necessarily understand who God is any better than a suffering homeless person or a street vendor who recognizes something of God in the basic goodness of people. However, it does seem to imply that concepts such as "moment of creation" and "cosmogenesis" are being seriously included in scientific musings, for scientists are focusing on what nature tells us about itself. Furthermore, certain scientists today are suggesting that, if humans are included in the term "nature," then natural scientists must include humans in their considerations of what "world" is.

Many people seem hesitant to acknowledge simplicity as a desirable quality. In order to value the mystery of simplicity, a growth in spirit is essential. We must come to applaud simplicity not as a deprivation, but rather as a perfection, a mark of genius, a conservation of a sort, a manifestation of plenty in but a little. Assimilation of this concept requires much time and reflection.

With regard to discoveries of the deeper mysteries of nature—the mystery of the Big Bang, in particular—renowned physicist John Wheeler once remarked, "I think the secret will be so beautiful, so compelling, that we will all say to each other, 'Oh, how could it have been otherwise? And how could we have been so stupid for so long?'"[4]

Perhaps the Grand Unification Theory, if ever one is formulated, will be based on something far different from what we may uncover using super-collider "desertrons." Perhaps, too, the simplicity for which we search is already staring us in the face, and we are simply unable to recognize it. Perhaps, ultimately, the greater need is for discernment rather than experiment.

Notes

1. For a short time, the discovery of helium in Texas gave the United States a strategic advantage over the Germans, who wanted to use non-flammable helium instead of hydrogen in their zeppelins.

2. Richard Feynman, *The Character of Physical Law*, BBC (1965), 34. This book contains writings from the sound tracks of a series of lectures by the physicist Richard Feynman, a Nobel laureate and professed atheist.

3. Thomas Aquinas, O.P., *Summa Theologica*, Ia, XIII, 12.

4. Denis Brian, *Genius Talk: Conversations with Nobel Scientists and Other Luminaries*, New York: Plenum Press (1995), 135.

For Discussion

1. Cite something from your life that has proven to be of far-reaching consequence, beyond what you had ever expected.

2. What are some of the "lengthy threads" that make up the "tapestry" of your life?

3. Drawing on the notion of "finding God in the atom," give an example of how you find God in a particular thing or situation.

CHAPTER 19

Does Chance Have Dignity?

"Coincidence is the pseudonym God uses
when he chooses to remain anonymous."
—Anatole France

The subject we will now consider is familiar in everyone's life, namely chance and accident. We commonly speak of "taking our chances" throughout life while trying very hard to "turn the odds" in our favor. In a speculative mood, let us now view the concepts of chance and accident from a viewpoint that may differ from what we have hitherto assumed about these topics.

Among the terms we use when speaking of chance are words such as *fortune, luck, coincidence, serendipity, godsend,* and *windfall.* The ways in which we use these words often betray our basic philosophy of life. For example, a religious person might prefer to speak of chance and accident in terms of "godsend." Religious persons who are highly sensitive to coincidence often visualize it as higher in meaning than what they might call "mere chance." Scientists, on the other hand, enjoy speaking of chance, probability, coincidence, and chaotic behavior in highly mathematical ways, which, of course, we will avoid here.

In the overall scheme of nature there is a very strong tendency for things to run down, to move from order toward disorder. Scientists speak of this in terms of *entropy increase,* with entropy signifying disorder. Living things, however, seem to constitute an exception to this tendency. The highly structured cells of our bodies, for example, arise out of disorder, such as from the squashy foods that we chew, thereby demonstrating something of a counter-current in the overall behavior of nature.

In contrast to an atheistic view that pictures life as a bizarre disturbance in the expected flow toward disorder, Pierre Teilhard de Chardin viewed the oppositely flowing currents of living and non-living things as two faces of a single overall behavior of nature converging toward what

he called the "Omega point." Views such as his consistently suggest cos-
mogenetic wonder of the highest kind ultimately arising out of chance
encounter starting with atoms interacting with atoms. Teilhard viewed
the phenomenon of chance as something of sacred significance.

Many religious writers attempt to bypass the issue of chance and accident when describing the creative action of God. They seem to say
that chance is of low significance, implying that an intelligent deity
would not utilize phenomena that display randomness as a mode of creating, especially in the creation of humans.

It may be that we often view God's creative engagement too much
in the manner in which we look at our own roles in life. We may properly think that, compared to using one's talents in methodical ways that
guarantee specific outcomes, gambling is a less rational way of earning
one's living. Accordingly, and perhaps erroneously, we may expect God
to exhibit creativity in ways other than through chance phenomena.
The idea of humans simply happening as normal outgrowths of nature
appears to worry many good persons. They seem unable to visualize God
as endorsing the workings of nature in a way that makes holy its ordinary
processes—processes that turn out to be God's own workings. And for
things of high significance to simply happen as if coming from nature's
ordinary ways of behavior, they seem to feel that such things cannot be
of God.

People of religious faith sometimes forget that, since we do indeed
exist, God did in fact intend to include us in the realm of existence. But
the word *intend* means something different in relation to God than it
does in relation to humans. Whatever it is that God intends *does necessarily happen*, whereas much of what people intend is only tentative and
does not necessarily happen.

Is it possible that humans came into the world by way of chance-
encounter phenomena intended by God as a loving mode of creation?
Might it be that chance and accident are recognizable as having dignity?
As harboring an unrecognized holiness?

More questions arise: What about God engaging in chance phenomena? Isn't such a notion the very thing that Albert Einstein rejected in
exclaiming that God does not play with dice? If God is love, how can
love find expression by way of chance? How can things of higher meaning come into the world by way of what is often judged (or misjudged) to
be of lower status, namely chance?

In order to wrestle with this enigma, let us imagine flipping a coin. It
will, of course, land either heads up or tails up. No one can predict with
certainty the outcome of that so-called random event. Were we, however, to flip the coin 1,000 times, we could then be quite certain that it
would land heads up about 500 times, and tails up about 500 times. The

more flips, the more order we recognize in the overall outcome, and the more accurate can be our predictions of outcome. Such behavior in nature is described by a principle of statistics that says:

> An individual random event is unpredictable, yet the overall outcome of a large number of random events is very predictable.

Let us look at this very important principle as expressive of an intrinsic hope residing within the inner workings of nature. It unveils a basis for suspecting that something of higher significance might arise from the sum total of repetitions and aggregations of things throughout nature...things that, taken individually, would appear aimless. Thus, the principle suggests a dignity arising from multiplicities of things regarded as random. This implies that, as we back away from apparently disconnected individual random events of various types, we are able to discern in them patterns of regularity leading to recognizable order if, indeed, order is what it takes to awaken our recognitions of meaning.

Now we are drawn to consider whether or not chance and accident have meanings more closely related to our basic humanity.

Two persons from cities far apart meet by chance and fall in love. Not long afterwards, they reinterpret the series of events that brought them together. They finally speak of their accidental meeting in more meaningful terms, as having been "born in heaven." And yet, from the viewpoint of science, they met by chance. Their meeting was a chance-event that was afterwards crowned with special meaning by way of their love. With that in mind, let us now consider a still more fundamental example of people praising God for highly transcendent creations that are obviously rooted in chance.

In the remote past, men and women thought of their gender as having originated from direct creative decisions of God who, as we might say, "made us men," or "made us women." Later, through the use of microscopes, biologists came to understand that one's individual gender arises from the particular type of sperm cell—one carrying an X chromosome or one carrying a Y chromosome—that meets the ovum in a fertilization process. Utilizing the ordinary language of statistics, we might now intelligently assert that it was approximately a 50-50 chance that determined the gender of every human in the world. In effect, it was as if God flipped a coin when deciding whether to bring about a man or a woman. Nevertheless, we still recognize the hand of God in the outcome, do we not? In response to that mystery people are heard to exclaim with elation, "God made me male!" And songs are written, such as the well-known "Thank Heaven for Little Girls!" Examples such as these reveal that chance phenomena might, indeed, be the basis for creations that we understand as wonderful.

Now let us extend our thinking another step further. If our sexuality is recognized not only as having come from the Lord, but also as an outcome of a chance-encounter phenomenon, then perhaps *all* of creation can be validly understood in the same way—as having come from the Lord through chance-encounter phenomena. Is it possible, then, that chance-encounter stands at the root of cosmogenesis? Who knows? However, one can think about it and wonder if it might be proper to regard it as a holy scheme utilized by divine Love to create the wonders of the cosmos.

It is not unknown for nature to behave in such ways. That is, if part of it manifests a certain very fundamental characteristic, then *all of it* seems in some way to manifest that same fundamental characteristic.

Why our fixation on the phenomenon of order? Cannot great things be recognized also in disorder? It is of interest that in the field of metallurgy, asymmetry in metal crystals greatly increases the strength of a metal. Polycrystalline copper, for example, is much more rigid than single-crystal copper, which has highly ordered structure. Also, in the manufacture of transistors, certain local imperfections are known to enhance the ability of semiconductors to function in better ways. In cases such as these, one has good reason to acclaim disorder over order.

Many of us have long delighted in recognizing orderliness as the work of an Intelligent Designer. Indeed, much good has come from this view. But there are problems connected with such an outlook. It implies that the phenomenon of disorder is necessarily lower in dignity than that of order, and that molecular chaos, like human-instigated chaos, discloses lack of intention or intelligence. There has been a growing realization, however, that the reverse might be true.

Scientists today are searching to understand nature's laws of chaos, while recognizing the term *order* to be inclusive of much more than previously thought. A great deal has been published regarding these recent scientific investigations, and new definitions of order and disorder are being proposed.

Meanwhile, certain philosophers of religion are suggesting that chaos, if reexamined, might be understood as revealing infinite foresight on the part of God. Disorder, they say, may have unrecognized higher meanings, such as "mother of order." Thus, one is led to surmise that what we have heretofore understood to be disorder might also be viewed as order of a greater complexity.

It is strange that we tend to place exclusive high value on our understanding of orderly arrangement when we ourselves often instigate disorder to bring about wonderful things in this world. Cannot God engage in planned disorder, let us say, in the manner of a chef mixing the ingredients of a fruitcake? If all of the raisins, or dates, or pecans fell together in parallel alignment like sticks of firewood neatly arranged on a woodpile, would we not consider the cake a failure? The randomness of locations of fruit in a fruitcake is an example of an instance where disorder is applauded. Cannot our understanding of nature's disorder be extended to include the concept of planned disorder having a place among the unfathomable ways of the Creator?

A further example of recognized order arising from disorder lies in the trillions of rapidly moving air molecules in an automobile tire. As they bump around individually and randomly against the inner walls of the tires, they deliver individual jabs against the walls. Trillions of these impulses add up to what we, as outside observers, call "steady tire pressure." And most of us, of course, stake our lives on this orderly outcome while driving at high speeds.

Persons who visualize the world in a highly transcendent way might well picture the countless particles in a dust cloud and say that, by God's providence, something of promise is about to emerge. It is well known that particles of atmospheric dust in chaotic motion from volcanoes in Italy help us to witness beautiful sunsets. Atmospheric dust also enriches the blue that is often seen above the Mediterranean.

Religious thinkers today are moving beyond the traditional maxim "Where there is order and design, there also is the Designer, the God of Order." Their visualizations are expanding to include a Designer much more intelligent than we once imagined, for chaos is increasingly understood today as pre-ordained order. God is being pictured anew as if personally focusing attention on every atom everywhere at every moment. And, being infinite, the Creator knows the outcome of all things for all time. Nothing can be accidental in the sight of the Lord. What we call "accidental outcome" is accidental only to ourselves, not to God. As scripture says:

> How weighty to me are your thoughts, O God!
> > How vast is the sum of them!
> I try to count them—they are more than the sand;
> > I come to the end—I am still with you. (Psalm 139:17–18)

An important scripture-related question arises: If humans came into this world along the avenue of evolution through chance-encounter phenomena—beginning with atoms and photons meeting one another—can a person today be a "chosen one" of God in the scriptural sense? The answer, of course, is "Absolutely yes!" as we will see.

The term "chosen one" as used in scripture refers primarily to the order of grace. So-called *chosen* persons perceive themselves as favored in special ways, as being adorned or graced. We who enjoy the gift of religious faith understand that it was granted to us; it was hardly something we could have earned. We understand that favor somehow came upon us as if from beyond ourselves. But regardless of how it came, the end effect was a love-awareness of ourselves as having been chosen by the Lord.

Believers know that, in the scriptural passages, God has revealed—often through the poetic or metaphorical expression of inspired writers—that we are chosen, that we are special, that our names are written on the palm of God's hand (Isaiah 49:16). Such discernment gives us an awareness that we are unspeakably fortunate to be God's friends. Yet our awareness of having been chosen by God does not exclude the possibility of that chosenness having been brought about by events based on random selection, survival of the fittest, and interactions that may seem arbitrary to us. Above all, knowing ourselves as having come into this world by chance phenomena is not incompatible with knowing ourselves as loved by God, for the ways in which the Lord can love are without limit.

In the eyes of persons of faith, all phenomena are God-expressions. No phenomenon is unholy in the sight of the Lord, who sustains all that exists. God is also the King of Everything, including random events, chance phenomena, and that which we commonly know as arbitrary, variable, irregular, capricious, undependable, and aimless. All of these are understandable as precursors of hope leading toward life in its broadest sense inasmuch as they rest in the God of Creation who is Life Itself.

God is the Lord of all that is, disorder included. God is the One who forever rescues, recovers, ransoms, redeems, reclaims, and restores. This does not necessarily mean that God actively rescues creatures in the way that we commonly think of the term "rescue." Suffice it to say that whoever is conscious of God at every turn is already rescued or delivered from the horrors of a world visualized as meaningless.

Most Christians do in fact see the providence of God in ways involving chance encounter. Obvious examples include finding a sum of money, winning a raffle, or unexpectedly bumping into an old friend—events to which we sometimes respond by exclaiming, "Wasn't it providential that...?" But how do we grow in our understanding of God's loving concern expressing itself in random selection, a term often considered in evolution? Let us consider the following example:

Suppose we visit an orphanage in Korea with the intention of adopting a child. We stand there looking at the yard where a hundred orphans are at play. For one reason or another, we suddenly, almost capriciously, make an agreement with the director that we will adopt the very first child who sneezes. As the outcome of that unusual agreement, we acquire guardianship of Kim, a three-year-old boy. We welcome Kim into our home and lavish him with attention, sharing our time and possessions with him, teaching him everything good as years go by. There is no denying that Kim was chosen through a plan of random selection, and yet our love is very obvious. When Kim matures, he will probably thank us for having chosen him over others. He need not ever know what determined our choice, but should he find out, that would be all right as well. For the overriding issue is that he understand that he was, and still is, personally loved by us. After all, the experience of being loved by another is very much one of joy over having been chosen and cherished, regardless of the details of the selection process.

It is difficult for us as humans to understand how love can be present in unpredictables that stretch back to antiquity. God's love for us is at work in everything, including chance-encounter phenomena dating to the earliest beginnings, even before the slime of the earth mentioned in Genesis. With God there is no such thing as "mere chance." God's love extends back even to the first moments of the creation of the cosmos, and to what immediately followed, a time when all matter was mostly hydrogen. And because it existed, it was therefore of wonder.

Because God is present everywhere, God's intelligent intent and loving concern for each and every atom and molecule was fully present in the countless random events and chance encounters that ultimately gave rise to and promoted the orderly phenomenon of biological life. Evolution itself is a drama of emergence in which matter, *through its ordinary behavior*, can eventually attain higher states of existence. Each and every particle, through its native tendencies, expresses in its own proper way its Maker who is holy, true, loving, necessary, absolute, and life promoting.

We ourselves can be insensitive about creatures. In place of the truthful statement, "I am unable to fathom their purposes and meanings," we easily communicate negative judgmental conclusions. One way we do this is by uttering words of calm disdain such as "mere," or "only," or "nothing but" when speaking about things. It is easy to dismiss as inferior that which seems of little use to us, or that which numerically overwhelms us.

Among the admirable qualities of scientists is that they search for ways to explain what is difficult to describe. For example, when they are unable to describe the individual activities of countless molecules comprising a gas, scientists revert to statistical descriptions of their average or "effective" behaviors. Each molecule of a gas is viewed as important, as contributing its individual mass and velocity to determining the overall state of the gas, including its measured pressure and temperature. In physics, the kinetic theory of gases is quite impressive, as it describes the external characteristics of gases in terms of the collective behaviors of their unseen individual, identical, and fast-moving molecules.[1]

Understanding that the term "important" is a highly relative one, we ought always to be careful about dismissing certain bits of creation as low in importance. Christians, in particular, are conscious of the spiritual risks involved in thinking negatively, for in the background of our musings, there stands the example of Christ, who was judged worthy of death.

It should neither surprise nor disappoint Christians to learn someday that the Lord God has for eons been calling forth something of highest significance—life in its various forms—from "mere" matter, much of which has traditionally been regarded as lowly. Specifically, we can reference inanimate atoms. Atoms, to their everlasting credit, go about faithfully obeying the so-called laws of nature, endlessly responding to whatever they are supposed to be and do. Who among us can be found to empathize with them, to appreciate them as they move to higher complexities, bonding in specific ways that bring about something of greater significance than before? Who among us can delight in molecules, too, as they move down the scales of complexity while decomposing into simple atoms? Few among us—scientists excepted—are appreciative enough to focus on them and applaud them for behaving everywhere in meaningful ways as they do.

Our vision is often not expansive enough to see creation's behaviors in the light of higher meanings, to turn our understanding into acclamation. Nevertheless, it stands as one of the glories of the scientific community that we include the very small in our thinking. In this way we indeed make the world more conscious of the wonder in things. When it comes to appreciating things, alas, small children are sometimes ahead of adults. Not surprisingly, then, scientists are sometimes chided by their friends as being like "big kids" having fun with their toys.

In science we pay attention to the so-called "aimless," discussing randomly moving molecules and theorizing about them. We correlate them and find meaningful ways to describe what they do. As if in return, they enable us to know ourselves as searchers, as persons who are in on their secrets as they prance around in the dance of disorder that sometimes gives birth to order.

From the viewpoint of creation spirituality, then, chance and accident can be recognized as highly meaningful. Like order, disorder too has grand significance. Indeed, certain philosophers have suggested that behavior based on probability *is an intrinsic property of nature* at some level.

A final word: What is written in this chapter ought not to be interpreted as a definitive and proven account of evolutionary development, but rather, as an invitation to keep open the doors of our thinking. Teilhard de Chardin viewed disorder on the primordial level as one stage in a process that leads by way of increasing "complexification" to life on earth, then to the appearance of humans, and ultimately to the salvific presence of God as one of us in this world.

Note

1. We refer to gases in this chapter, reserving reference to solids for a later chapter that will focus on geometrically perfect solids arising from randomly moving molecules slowing down to cohere and form crystals.

For Discussion

1. Can you cite an example or two where unpredictability in your life has been desirable?

2. What makes some things "ordinary" and other things "special"?

3. Could someone who is widely regarded as "ordinary" also be "special"?

CHAPTER 20

Matter and Moral Consciousness

"The proper thanks [for creation] is some form of humility and restraint.
We should thank God for beer and burgundy by not drinking too much of them."
—G. K. Chesterton

Very much of what is written in scripture, particularly in the Old Testament, is couched in terms of cosmic phenomena—deluge, earthquake, burning bush, pillar of fire—mystically understood. Prophets of old, touched by grace and listening from the very core of their being to the insinuations of nature, acquired new insights into proper behavior. They regarded these lessons as coming to them from Yahweh, their one and only Lord God.

It should not be difficult for us to imagine the voice of God being fathomed by the prophets of old. Touched by grace as they viewed creation with receptive hearts, they saw so much more than met their eyes. Through the medium of creation they sensed what lay beyond things living and non-living—something of profound presence, favor, truth, wonder, and love. Recognizing creation's majestic meanings, the prophets voiced their convictions with an urgency amounting to intolerance against complacency and business-as-usual attitudes, for they understood creation's meanings to be as genuine as life itself. Their openness and response to the transcendent meaning of creation suggest that, if a Divine Will is somehow embedded in creation from the beginning, it lives in a special way in the human psyche. And it is especially present to those who are receptive to meaning, humble individuals who prayerfully follow their star.

Advancing now to modern times: If we acknowledge that God works from within creation, we should include the hidden world of subatomic matter when considering the world's role in promoting moral consciousness. This hidden world has certain strange aspects, especially in the case of radioactive matter, for there is something highly unusual about radioactive materials, especially those with *fissionable* atoms.

Fissionable atoms, when placed in close proximity to others of identical makeup, trigger one another into disintegration. For example, when a single neutron strikes the nucleus of a uranium-235 atom, that atom breaks apart into two new atoms—one of krypton and one of barium—and *three additional neutrons*, each of which is then capable of triggering other U-235 atoms into disintegration (see illustrations below). The disintegration process is a multiplying one, the overall outcome being the now-well-known chain reaction of nuclear explosions.

For a chain reaction to occur, sufficient atoms must be crowded closely together to create an effective target area for subsequent collisions. It may be helpful to consider an analogy. Let us imagine a room in which there is no furniture. Imagine further that we place a dozen mousetraps here and there on its floor. Each mousetrap is then set, with three marbles being carefully loaded onto each spring. Imagine that we then throw a single marble into the air above the mousetraps. As our marble falls, what are the chances of all twelve mousetraps being set off by our single marble? Almost zero. Why? Because the mousetraps are far apart, thus presenting a very small target area for a marble thrown into the air.

Let us now repeat the experiment, but after covering the entire floor with, say, ten thousand mousetraps. Each one has been set and loaded with three marbles on top of it. What would now happen if we throw a

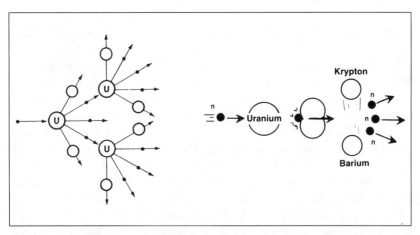

Illustrations showing the principle of chain reactions. A single neutron from the left causes disintegration of a U-235 atom into krypton and barium atoms and an additional three neutrons, which in turn cause neighboring U-235 atoms to disintegrate in multiplying fashion. The important point is that one neutron coming in results in three going out.

single marble into the air above the traps? Answer: All of the mousetraps would snap, because the target area has been vastly increased. A single marble hitting one mousetrap would cause it to snap, releasing three more marbles. Each of these, upon hitting another mousetrap, would cause it to snap and release three more marbles, thus placing nine marbles in flight. Each of the nine, upon striking a trap, would cause its marbles to fly, and so forth, until all ten thousand mousetraps will have been set off, throwing thirty thousand marbles up in one gigantic chain reaction. Figuratively, then, the minimum number of mousetraps we would have to place on the floor in order for this chain reaction to occur is what is signified by the term *critical mass*. In terms of radioactive materials, critical mass refers to the minimum amount of a radioactive fissionable material that is needed if a chain reaction explosion is to occur. Subcritical mass refers to a mass of fissionable material that has less than critical mass.

The notions of critical and subcritical mass raise some philosophical and religious questions regarding the ways that we sometimes erroneously think about nature's behavior. Let us see how these two concepts can help shape our moral consciousness.

From a spiritual perspective, it is often easy to equate the concepts of togetherness and harmony with the so-called "coming of God's kingdom." We normally visualize harmony in terms of people uniting with a common interest, at a convention of bird watchers, for example. But when citing harmony in creation as a whole, we must be cautious.

Many educated people today do not realize that, to initiate a nuclear explosion, we do not need a match, a fuse, or an electrical spark to detonate nuclear devices. We need only bring together two subcritical masses of fissionable material. In the case of a uranium bomb, this means bringing together two masses of uranium isotope 235, each being about the size and shape of half a grapefruit. As they are brought together, critical mass is suddenly reached or exceeded. The ensuing fast reaction initiates an explosive release of tremendous energies.

But how can this be? Would this not mean that U-235 atoms, as found in nature, could automatically trigger one another into explosion? The answer is no, and the reason is relatively simple.

As found in nature, uranium consists of only one atom of U-235 to 140 atoms of U-238. That is to say, in their native condition, fissionable U-235 atoms are lost in crowds of non-fissionable U-238 atoms. Being so separated from one another, they do not significantly interact with their own kind. If they are to react, U-235 atoms must be extracted from naturally occurring uranium and then concentrated, as was done in the uranium enrichment efforts of the United States during World War II.

In pondering these things to discern their implications, we might ask: What is U-235 telling us about itself? It is not difficult to imagine that U-235 atoms, if they could speak, might say the following:

"Nature intends for us to remain far apart from the rest of our kind. Unless you keep us separated, expect us to interact and destroy each other. Thou shalt not exceed critical mass!"

If we were to translate such an unusual message into the realm of humans relating with humans, we could use words such as these:

"I certainly am all right as I stand here alone with you. Calm as I am at the moment, I must advise you not to invite too many people like myself into your home at one time. For we would inevitably irritate one another and instigate terrible fights. We would likely come to blows and leave your home in shambles. It is in our nature to do this, and we cannot behave otherwise!"

Philosophically speaking, the implications of the U-235 message are unlike those of everyday matter that we engineer into specific shapes and forms. They differ sharply from the understanding we garner from steel, brass, or lead shaped into handguns, cartridges, and bullets. The essential difference is that the U-235 message is primarily one of substance, not of form. It proceeds from the fact that the 92 protons in the nucleus of uranium atoms constitute a loosely bound entity, stuck together yet ready to fly apart. It is a strange message because we habitually imagine birds of a feather flocking together in peaceful accord. But, in this case, we must acknowledge that some identical creatures in this world are by nature incompatible with each other, unable to endure physical intimacy with others of their own kind.

Therefore, in the majestic language of transcendence, and with ages of time in mind, people of faith might meaningfully speak as follows:

"In creating uranium-235, God placed (or allowed) within its atoms a behavioral propensity from which a unique message would one day come forth. In time, it would be delivered to rational creatures on a planet destined to be called Earth. The behavior of these atoms would one day jolt the moral sensitivities of humans."

Messages derived from the behavior of U-235 and other fissionable materials are presently being discerned with universal seriousness. The manner in which God implanted, sanctioned, or endorsed the behavior of fissionable matter is, of course, open to scientific, philosophical, and theological musings. We might think that, since radioactivity is rooted in random disintegrations among certain atoms, the U-235 message cannot be of God. This, however, is not true.

Inasmuch as creation is basically good, phenomena of chance, accident, randomness, and disorder are no less sacred than other behaviors of nature. As was explained in the preceding chapter, disorder is frequently at the root of order in things that we know as steady, dependable, and aesthetically uplifting.

To an astonishing degree, the secrets of how matter-energy and space-time originally came together are tumbling into the hands of scientists today. Meaningful understanding of life's building blocks—DNA configurations, living cells with structures derived from encoded messages, for example—is now entering human awareness. Astounding answers are coming forth from the simple question "What is matter?" And many of these answers raise further questions as to what constitutes life, and how life ought to be treated.

Nature itself—or God at work through nature—is compelling us to look at "plain old matter" and to see it anew. Strangely, a foremost moral imperative of our day—not to unleash nuclear warfare—is one arising from humankind's understanding of the inner workings of extremely tiny atoms.

With the explosion of the first atomic bomb, people everywhere were shocked into awareness of the stupendous energies concealed in small amounts of matter. To this day, however, most people are unaware that no extraordinary behavior of matter was at work in the atomic bomb. What was at work was the very ordinary, normal behavior of radioactive U-235. It was the "same old matter," but in the unusual condition of having had a concentrated abundance of its atoms brought together in close proximity.

The destructive power of atomic explosions caused people everywhere to look at material substance as they had never done before. It was as if matter itself, just by being what it always has been, thundered a powerful ethical verbalization, one that even the most insensitive would hear with understanding. At long last, the unseen inner workings of the world gained their place in the consciousness of people. Like Christ the eternal Word, energized matter awaited the fullness of time to be heard in history. And now, having been heard, it cannot be ignored without risking extreme repercussions for the entire human family.

Christians would find it difficult to accept the idea of material substance and circumstance promoting moral consciousness on their own, as if apart from God. There is a very good reason for this. After all, when speaking of the world from the Christian perspective, the concept of "apart from God" is meaningless. If God is holding everything everywhere in existence, the concept of "without divine concurrence" has no real meaning.

For Discussion

1. To what extent do you visualize yourself as able to do things "strictly on your own" as if aside from God?

2. Name one or more possible good uses that could be made of a nuclear bomb.

CHAPTER 21

On Science and Evolution

*"In its own way, matter has obeyed from the beginning
that great law of biology to which we shall have to refer
time and time again, the law of 'complexification.'"*
—Pierre Teilhard de Chardin

An article in the *New York Times* some years ago began with this noteworthy statement: "These are interesting times to be in physics. There is speculation that both the history and the fate of the universe is written in every atom."

Statements such as this certainly attract attention. Shortly afterwards, an article entitled "The Life of the Mind" appeared in *Notre Dame Magazine*. It explained that scientists for the first time were beginning to understand *on a molecular basis* how the human mind works. Among researchers on the human brain, the term *chemistry of awareness* is coming into use. One article describes the new field of *chemical epistemology*, wherein studies based on chemistry are being conducted in an effort to determine how we come to know what we know. Although exciting, thoughts like these seem to worry many good persons who believe that their religious beliefs are somehow at risk.

In the present-day intellectual climate, it is difficult to dwell on meaning in the sciences without touching in some way on the subjects of creation and evolution. To the extent that we appreciate living, we tend also to view the origins of the universe and of human life as having super-special significance. We yearn for answers to the question of how the human race began. We fantasize about how we would have delighted in having been "on the scene" as witnesses to the stupendous event of creation's beginning! But we easily forget that such fantasies might involve contradictions. For example, how could a person, by nature a part of the cosmos, witness a transition of the cosmos from non-existence to existence? To witness such an event with understanding would require human intelligence at a time when humans and Earth had not yet appeared. It doesn't add up.

Opinions in favor of physical, chemical, and biological evolution run strong in the world's science communities. Examples of present-day evolutionary processes in physics, geophysics, and cosmology include: the transmutation of certain chemical elements into other elements by radioactive decay, the increase in entropy (disorderliness) of the universe as a whole, the shift in location of magnetic north, the drift of continents due to oozing substances from the earth's interior, and the formation of stars from proto-stars.

From the viewpoint of scientific evidence gathered thus far, how did our evolving cosmos begin? One approach to an answer, of course, is to feed into our computers all significant accumulated data on the present state of the cosmos, along with all the known physical laws that express cosmic behavior and that have brought us to our present state.

Combining the basic data gathered from several fields of science, cosmologists calculate that everything in our vast cosmos originated from a super-dense entity billions of times smaller in size than what we presently understand as a proton! Incredible! Their computers calculate that the primordial "Big Bang" explosion, marking what is tentatively regarded as creation's beginning, occurred about fifteen billion years ago during a time interval of about 10^{-43} second! The implication, of course, is that everything in the cosmos today stems from that single super-colossal event, a thought that truly boggles the imagination![1]

As to the evolution of the chemical elements, cosmologists believe that the hundred or so kinds of atoms that we know today originated more or less in the following way:

Hydrogen was formed during the first minute or two after the Big Bang, the colossal explosion that caused the boundaries of space itself to expand. Later drawn together in localized clusters by tremendous forces of gravitation, the hydrogen atoms coalesced into giant *blue stars*, nuclear furnaces supportive of the fusion process in which hydrogen nuclei merge and become helium. As the hydrogen in the stars became less plentiful, atoms such as carbon, oxygen, and sodium were formed as a kind of ash from hydrogen and helium. Iron and the heavier, less plentiful elements, such as gold and uranium, were produced by super-squeezing, with nuclei touching nuclei in the blue giants that finally contracted to explode as supernovae.

These materials from exploded supernovae were hurled into space, only to gravitationally coalesce again in *second-generation aggregations*. It was from these aggregations that solar systems like ours were born. The much-photographed Crab Nebula, six thousand light-years from the earth, is the outcome of a supernova explosion observed and recorded in the year AD 1054 by Chinese astrologers. In such ways, the wonderfully different materials, including those presently in our own bodies, are believed to have originated deep within the stars and then to have been disseminated through space by star explosions.

The elements listed in our periodic table of the chemical elements are believed to have been an outcome of this process. In our own day, scientists have artificially produced new kinds of atoms by bombarding known atoms with high-velocity particles. This leads us to believe that the more complex heavier atoms evolved from simple hydrogen atoms. Thinking along this vein, then, it would be of little poetic exaggeration to say that our planet itself and everything on it, including our own bodies, was formed from "star dust."

Our solar system was almost certainly formed by materials from exploded stars, the remnants being drawn together by gravitation and coalescing off-center so as to produce a hydrogen core with whirling materials surrounding it. At the center of this system, our sun underwent formation. In accord with the behavior now described by Kepler's third law of planetary motion (see pages 51–52), the whirling materials nearer the blazing core had to separate from the slower-moving materials farther away from the core. These slower-moving materials, drawn together under their local gravitational forces, formed into rotating and orbiting planets. Material orbiting the planets coalesced to form moons. Our present-day infrared telescopes are photographing planets being

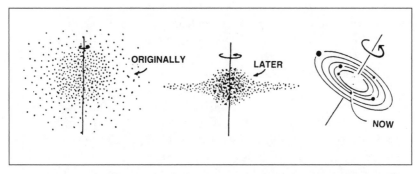

A sketch illustrating how materials being drawn together by gravitation coalesce into a whirling system, with some of the materials orbiting the central core. It is of interest to note that the equator of our sun rotates in the same direction as our planets, and that the planes of the planetary orbits coincide with that of the sun's equator (with a slight discrepancy in the case of the farthest and least tightly bound planet, Pluto).

formed in just this way around the stars Vega and Beta Pictoris, further reinforcing our understanding of how our solar system evolved.

During the 1950s there was speculation that primitive life on earth consisted of a kind of carbon-based scum. The presumption was that in the course of time, on a planet having the right conditions, linkups between carbon atoms would naturally occur. Eventually they would evolve into substances having molecules like those known to make up living things today. Then, during the 1960s and 1970s, carbon compounds were actually observed to exist in space itself. Amino-acid molecules were found on meteorites, suggesting (but, of course, not proving) that the basic components for life on earth *might have come* from outer space. Scientists then discovered that they, too, could produce amino-acid molecules by sparking methane gas, a substance known to have existed on primitive earth. Today there are known forms of organic clusters—viruses, in particular—that are so primitive that biologists debate whether they are living or non-living. Their very existence on the fringes of recognizable life suggests the possibility of living matter naturally evolving from non-living.

The molecular explanations of life's beginnings state that carbon atoms on a planet's surface can link up with other atoms by natural processes. In this way nucleic- and amino-acid molecules are thought to form. In moving about, they bond to one another in specific ways to form DNA chains, the atoms of which are comparable in number to stars in a galaxy. DNA chains possess the characteristics of a genetic code that is transferable and reproducible. When a double-helix strand of DNA "unzips," each half then becomes the basis for the DNA of an emerging daughter cell. Thus, the scientific community presently has strong suspicions that life on our planet has arisen from non-living matter by way of natural processes. Life is viewed as having sprung from blind encounters of atoms and photons, a process understood by some as totally impersonal and, therefore, discordant with religious views of creation. How unfortunate!

Some people seem to be disturbed by the idea of life "just happening" from what was already there. They view spontaneous generation as an activity that is devoid of intention, lacking evidence of love, and therefore unworthy of a Creator who is said to be Love itself. Today this issue is rapidly becoming a key point in the faith stance of many educated persons who are aware that a molecular explanation for life is already in our textbooks. It is referred to as "the chemistry of life." The important question, then, seems to be: Can we honestly say that the genesis of human life via so-called "accidental encounter" of things holds very high meaning?

Note

1. This so-called "Big Bang" theory ought not be understood as the only valid cosmological hypothesis for the world's beginning. Among other prominent hypotheses, for example, is one described in the November 1994 issue of *Scientific American* under the article title of: "The Self-Reproducing Inflationary Universe."

For Discussion

1. Are you disturbed by the suggestion that God might have created some or all living things by way of a gradual progression from non-living to living things?

2. If so, why?

CHAPTER 22

Evolution as God at Work in This World

I t is unfortunate that some persons, in wishing to take a stand on the side of God, view science as opposed to religion. They seem unable to entertain the possibility that living matter might have come forth, or might some day be made to come forth, from interactions of non-living materials. They seem to assume that nothing as wonderful as life could possibly arise from "lowly" inanimate atoms and photons interacting with one another. They suppose that the concepts of chance encounter and chaos are devoid of intelligent intention and, therefore, far removed from the idea of God. In the view of this writer, their position seems to deny God the freedom to create life on earth by way of chance and accident should that be God's will.

Now, it is easy enough to say: "Scientists have achieved some wonderful things, but they will *never* accomplish the synthesis of life from non-living substance. Only God can create life!" This might seem at first to be highly affirming of God. Highly affirming—yes!—but also highly restrictive. Those who voice such statements essentially set up scientists and God as opponents, for they rule out the possibility that the actions of scientists can be interpreted as God's creative activity being accomplished through the minds and hands of humans. They imply that God must act alone and directly when creating life. This pious viewpoint, however, also implies that they have figured out God to the point of asserting that God would never create new life by means other than those we can already imagine.

On the other hand, it may be that what we commonly think of as God's "action" is eminently passive or highly unassertive from our ordinary point of view. It may be that God's power is not power in our conventional understanding of the term. It may be that God has created life from its beginnings by quietly allowing events to occur, one moment leading to another, one event giving rise to another in such a way that myriads of such events led to at least low forms of life appearing via a process in which non-living matter was energized into living matter. This view of creation would not be out of tune with Christian thinking if we recall how scripture itself shows God's indescribably great influence at work through almost unbelievable events, such as the Father allowing the crucifixion of his Son.

Many religious thinkers believe that God accomplishes things in today's world by way of specific tendencies implanted in creation from its very beginnings. They think that great good has occurred across the ages just by things being (what seems to be) passively allowed to develop according to those originally implanted inclinations. They imagine this to be the manner in which the substances and situations of the world came to exist as they now exist, interacting and behaving within their ordinary fields of potencies and modes of demeanor.

We should note that reflections of this sort do not deny the poetic expression in Genesis of the grandeur of creation's beginnings. They simply portray the beginning of creation as being more compatible with the way nature appears to us today—with us as its listeners, and with God communicating to us through our growing understanding of nature's behavior.

We still we might ask: "How can life ever be understood as having arisen from non-living things? Isn't this a bit bizarre?" To answer this, let us first focus on a basic phenomenon such as crystal growing, in which order is clearly known to arise out of disorder. Consider a copper sulfate solution with its trillions of randomly dancing molecules. On cooling, these molecules slow down, lock together in discrete ways, and form well-structured geometrical crystalline solids. This phenomenon of freezing exemplifies structured molecular arrangement arising out of disorderly molecular motions within liquids.

We can picture the disorder-turning-into-order arising in crystal growing as a kind of short-range evolution in form, with liquids evolving into solids across short spans of time. Indeed, there is speculation today that crystals, because of their ability to reproduce their forms and grow, may be the immediate pre-life forms in our world.

Proceeding further, we can also view chemical changes as a kind of short-range evolution in the makeup of substances wherein two or more substances, on interacting, evolve into a third substance. Chemists visualize the "seeds" or potencies for chemical changes as residing within the atoms themselves. Accordingly, they predict outcomes of chemical reactions in terms of such things as valences, energy levels, and laws of constant proportions.[1]

The tendency for atoms to interact in specific and unique ways as if in obedience is very well known. The outside views we have of crystals—

the geometrically ordered externals we see—are the result of inner workings that we cannot directly see, but about which we know a lot.

Snowflakes exemplify a particularly beautiful kind of crystal growth. It is interesting to note that the Book of Job asks the reader in poetic words:

> "Have you entered the storehouses of the snow, or have you seen
> the storehouses of the hail?" (Job 38:22)

Today these questions of wonder can be answered with a resounding "Yes!" for we now understand quite well how snowflakes are formed. We have studied how hailstones grow in layers after repeated vertical journeys through moist air currents encountered at higher and lower altitudes. We have, indeed, visited these places with our aircraft and rockets.

A century ago, many people, just like the authors of the Book of Job, pictured God as directly shaping the beautiful geometrical patterns of snowflakes. The pious thinking in those days was that no two snowflakes had ever been found to be identical because the Lord was actively at work, creating and directing the form of each snowflake individually in endless diversity beyond our understanding!

Realizing that snowflakes are products of nature, and without denying that nature itself is from God, we are led to ask: What known factors in nature's behavior account for snowflakes forming as they do? In partic-

A magnified view of snowflakes showing their spokes, identical on each flake yet different from the spokes of other snowflakes. Of special interest are the tiny gear-wheel-like configurations at the centers of the flakes where growth begins.

ular, knowing that the individual spokes on a given snowflake are identical, we may ask how atoms at the tip of one spoke "know" what's going on at the tips of the neighboring spokes so as to produce a flake having identical spokes? Over time we have uncovered the answer.

Snowflakes of endless variety, each exhibiting its own identical spokes, are formed by the natural tendency of water molecules to join together in particular ways as the temperature drops. The water molecules fit together by chemical bondings under the prevailing changing conditions.[2] And we must consider that a growing snowflake, while falling during its formation, undergoes identical environmental changes on all six spokes at once. This results in identical redirections of crystal growth (outward from the center) on all spokes as additional water molecules from the atmosphere pile on. Thus, identical spokes!

Thinking in spiritual ways, we might well view the written expression of such scientific knowledge as constituting books of revelation, for such writings describe in detail some of the wondrous ways in which the Lord accomplishes creation on a truly grand scale. From this perspective we come to see the tendencies within atoms and photons—such as the ones arising from their valences—as manifesting the Creator's faithful governance of creation from within matter. In assigning innate inclinations to the within of things, God can be understood as endorsing their behavior or (poetically, of course), as commissioning them to "take care of themselves" in God's name. By way of their valences, God can be understood to have appointed atoms to do what they do simply by being what they are. If ever we are to introduce and promote transcendent definitions in science, we would do well to include considerations such as these.

We might also imagine that, in a fully enlightened world, all correctly written science texts would one day be taken together and viewed in a spiritual context as a Second Book of Genesis. Some might understand these collective works of science in terms of principles that bring to light something of how God created the world. Others might see science books as illustrating the ways of the world and, by extension, the ways of the Lord who created the world and programmed it to go forth in time. Indeed, the ways of perceiving God through creation are endless.

Through renewed faith on the part of Christians, then, the written works of science might win recognition and respect from the viewpoint of God *having infused the call of grace into everything*. When we view the world from this perspective, we might perchance suspect that the original creation was endowed with a silent, sacred, and persistent mandate given to all things to behave as we presently know them to behave.

And, in accord with this view, scientists have been the ones who have discerned these consistent behaviors, classifying them within specific categories, giving them names—terms such as crystallization, condensation, evaporation, transmutation, polarization, mutation, and conservation. Thus, those who are fully at peace with the makeup of the world may view science in the light of God having assigned physical and chemical properties to substances. To atoms God would have assigned the task of expressing their vigor agelessly, with self-reflective humans, endowed with special blessings, eventually arising out of matter-energy at work within the realms of space and time considered as sacred.

Returning for a moment to crystals: Because of their geometric shapes and overall beauty, crystals are the kind of thing we easily recognize as extraordinary and somewhat fascinating. As freezing liquids turn into solids, the emerging geometric arrangements involving "the same old atoms as before" invite us to see them anew in terms of "beauty on the increase." Their orderly crystalline form invites us to recognize them as wondrous. Blue copper sulfate crystals are especially attractive.

Often the magic in things seems to vanish as we learn more about their workings, which, we sometimes say, "explain" their mysteries. On uncovering the mysteries of snowflakes, for example, one might exclaim, "Oh, now I see!" We might in fact grow to understand their molecular structure at the expense of losing touch with a mystery of even greater significance: the simple-yet-remarkable fact of snowflakes existing at all. Sometimes we forget that the greater wonder lies not in the scientific explanation of things, but rather in their being what they are in a reality where, compared to nothingness, existence itself is the ruling consideration. The true wondrousness of things is their very existence as they go about exemplifying countless modes of being that have come into the world, with each thing dramatizing in its own unique way a possibility made actual. Always, the being of a creature is the true mystery. The imaginative procedure of visualizing the being of creatures against a background of nothingness is an innately high engagement. Most of us do not engage enough in this kind of thinking.

Today's biological scientists are suggesting that potencies for life are harbored in non-living matter. They suppose that, in the vast majority of cases, the seeds of life in the non-living never sprout forth into life simply because the proper conditions for life are not ordinarily met, because living molecules are extremely complex. Now, contrary to what religious fundamentalists might say, this view of creative action is *not opposed* to a theistic interpretation. Rather, it presents us with the faith-challenge of putting aside old notions that inanimate matter is lowly and wanting in dignity. We so easily fall into habits of thinking that common things such as mud puddles, dust, and broken glass do not count for much because they give us trouble, get in our way, appear asymmetrical, soil our clothes, or emit unpleasant odors. But the challenge is to prayerfully modify our attitudes toward that which stands before us night and day— magnificent creation in all of its forms, ramifications, and implications. Outward expressions of the inner life of God!

In the event that biological scientists do, indeed, energize matter and obtain living cells from interactions of non-living matter (such as from clusters of atoms, amino-acid molecules, ultraviolet photons, and the like), this would not be reason to dismiss the concept of God as Creator and Author of Life. To do so would imply a belief that God could not have created beyond the ways we have traditionally imagined.

Is it conceivable that a recipe for life might one day be discovered, and that it might be identical to one from which at least some living things have arisen? Is it possible that the recipe might fall into the hands of a scientist this very day, regardless of whether he or she happens to be a Christian, Jew, Muslim, Buddhist, Hindu, or professed atheist? What if this does in fact one day occur? What if we were to read in tomorrow's newspaper: "Researchers Produce Living Matter from Non-Living Matter"? Were this to happen, how would it affect our perceptions of who God is? What adjustments in our thinking would be called for? To what extent would we be receptive to new possibilities in this day and age, when life forms are already being manipulated by way of genetic engineering? We have much to think about in this regard.

Notes

1. Those interested in the philosophical problem of "the one and the many" relative to crystals forming from cooling liquids would do well to read Michael Polanyi's *Personal Knowledge: Toward a Post-Critical Philosophy*, Harper and Row (1958), 394.

2. The number of spokes (six) is determined by the angle of the hydrogen-oxygen-hydrogen configuration within the water molecules.

For Discussion

1. What are your thoughts—uplifting or depressing—concerning the suggestion that all things in the world today might have come, as certain cosmologists insist, from hydrogen acting as it acts?

2. Do you think it is possible that an atheist might be the first to produce living matter from non-living matter? Were that to happen, would you be "disappointed" in God?

3. As was noted in this chapter, many religious thinkers believe that God accomplishes things in the world by way of specific tendencies implanted in creation from its very beginnings. If God were to have worked through amino acids, DNA, and great expanses of time to produce human beings, do you believe they would have less dignity than if God had produced them directly from "the slime of the earth"?

CHAPTER 23

Cosmic Yearnings

It has often been said that hope springs eternal. How true that is. In the world of the psychological, hope is the motivator and mover in even our most trivial actions. The simple act of turning our attention away from one thing and focusing it on another exemplifies an engagement in subliminal hope on our part. But what about hope throughout the realm of creatures in general, including even insects?

Close-up of a queen bee attended to by several workers. The caring attention given to the queen demonstrates a kind of yearning toward fulfillment on the part of the workers. (Photographed by the author).

Is there a kind of hope or yearning to be found in the realm of the non-living? Chemists experiment with matter, taking into account the affinities of atoms for one another. By applying known formulas and following the so-called "rules" of atomic behavior, they sometimes create unique molecules such as those of petroleum plastics. Chemists speak of atoms in terms of their binding energies, energy levels, valences, electron spins, potential gradients, and such. If questioned as to why atoms behave as they do, most would respond by saying, "Well, this is just the way they behave! What more can we say?"

But, indeed, there is much more that we can say about matter in dimensions beyond those expressed in our science textbooks. For example, we can describe how the components of the cosmos behave as if in response to deep-seated yearnings, exhibiting instincts of a sort. In exploring this, we shall depart for a moment from our usual manner of thinking and reflect on atoms in ways more gracious and benevolent than the manner in which they are described in our science books.

We can state with good reasons that the cosmos with its countless atoms and photons is a mind-boggling array of things that have been drawn together toward a "community" of a kind. Through gravitation, chemical bondings, electrical and nuclear attractions, and other unifying interactions, an overall "thrust" of some kind is discernible. Where aggregations result, things interrelate and respond to each other's presence so as to change their motions, such as when colliding, or to form new substances, such as when undergoing chemical changes. Might all of these atomic activities be validly understood, then, in terms of yearnings and hopes in search of fulfillment?

We can continue focusing on chemists, who deal in a special way with the expectancy of matter. When writing a chemical formula, they draw attention to cosmic hope-expression on a very fundamental level, for chemical formulas express what new substances *we can look forward to* when certain kinds of atoms interact in specific ways. Thus, we can think of chemical formulas as descriptive of the yearnings of atoms, whether these yearnings are characterized as attraction or repulsion, manifesting a steadfast readiness to respond to one another's allure in discrete ways, exhibiting reliable behaviors dating back to cosmic origins.

As if made for one another, atoms come together to form molecules. The outcomes of their interactions are very predictable in terms of their availabilities, their valences, their ways of bonding, and other factors. In fact, present-day materials throughout the cosmos can be described in terms of particles everywhere having consistently "followed the rules," such as when bonding together. For example, so much hydrogen when burned with so much oxygen will yield a particular and predictable amount of water.

But do they really follow rules? Or do they simply behave in marvelously consistent ways so as *to appear to us* as if they are obeying? Who can say? The fact is that matter itself, in spite of its seeming to be so common, remains a very great mystery. But let us assume at least for now that, unlike ourselves, atoms do not *consciously* follow rules.

When atoms encounter atoms, including those of a kind they have never met before, they always act as if they sense what they are "supposed" to do. This is the quality of atoms that especially endears them to chemists who lose themselves in the so-called "laws" of chemistry, a domain where atoms combine and separate in enchantingly precise ways.

People of faith often sense a kind of ongoing divine presence in all of creation, an intuition that must necessarily address and include the basic components of the world, namely atoms and molecules. Now, the exact ways in which these tiny building blocks are viewed may vary, depending on who is doing the visualizing. Some might envision chemical interactions in terms of begetting and achieving, picturing certain substances as uniting to beget new substances, such as sodium and chlorine chemically combining to "produce" sodium chloride, table salt.

This is not to suggest that chemists would think of God as mechanically pushing and pulling atoms toward specific ends, for God's governance of the world is infinitely more subtle than that. God can, however, be viewed as "acting" through things simply through the fact that things act according to their God-given ways. We can observe this in a science such as chemistry by imagining atoms as harboring inner yearnings reminiscent of instincts on levels proper to themselves. Dependably, they go about interacting, undaunted and doing what they characteristically do. Or, with purpose in mind, we might say, "doing what they were made to do."

We know that atoms of different kinds unite in discrete and consistent ways as expressed by chemical formulas. New molecules consisting of old atoms are formed. But this is not the final word in our understanding of matter.

If we think of creatures as being particular expressions of God, it then becomes possible for us to view them and their activities in highly quixotic yet practical ways. For we indeed have the freedom to think and speak of anything anywhere in terms of both science and religion—as exhibiting its physical and chemical properties and also as being a unique God-expression in that instance. It is unfortunate that many highly intelligent persons do not seem to exercise this freedom.

For one reason or another, people seem comfortable when speaking poetically of a beautiful sunset as an expression of God. On the other hand, they seem less than comfortable when speaking of a toothbrush, a screwdriver, or a burned out electric light bulb as an expression of God.

And so, with the help of science, how can we extend our understanding of things as God-expressions so as to include atoms and molecules and, consequently, everything everywhere? It is an ambitious task for sure, but one that we can initiate by interpreting the world from the viewpoint of interaction and interplay among God-expressions, one with the other. What might be some examples?

Let us first concentrate on creatures that are especially appealing—for instance, children. A child, herself a God-expression, who is tying her shoelaces can be viewed as engaging one God-expression (string) with another (eyelets) in a particular space-time God-expression (motion), all in hope of a finalization/revelation. A honeybee God-expression can be thought of as visiting with hope a flower God-expression in search of fulfillment involving nectar God-expression, the final outcome being honey-in-the-comb God-expression. These examples illustrate transcendent ways in which we can picture the visibles of the world interacting in an endlessly interwoven exhibition of God-expressions, one with the other. But what about the "withins" of matter, the domains that are not immediately visible to the eye?

Even here we shall see that the same kind of thinking is possible. For instance, those who understand chemistry might choose to visualize chemical reactions in a poetic way as follows:

Let us say that two substances are reacting to form a third. Thinking compassionately, we can visualize the original atoms as holding promise toward the emergence of new molecules in this world. We can also picture the emerging new materials as calling upon the old for their substance. In the microscopic world where newly formed molecules contain original atoms that have come together, we can speak of fulfillment as having occurred, or of a dormant hope within atoms having found fulfillment.

At this point one might ask, "Isn't all of this just a coloring of the facts, a mere configuration by the human imagination?" The question seems plausible. Our answer: "Yes, it certainly is a coloring of a sort. But the word 'just' or 'mere' is inappropriate in this regard, for the enhancing of nature and the bestowing of meaning on its makeup is exactly what the human mind is all about."

Aspirations and fulfillments throughout nature are occurring everywhere in endless ways. Aspirations are discernible in atoms combining as if meant for each other, in molecules disassociating or decomposing after having served their purpose and reverting back toward simple atoms; in atmospheric water molecules slowing down in their motions, adhering to one another and falling back to earth as rain; in excessive electrons within clouds yearning for protons on earth, a phenomenon that results in lightning from cloud to earth, or earth to cloud. When we think in gracious and benevolent ways, the scientific concept of "things doing what they do" translates into the religious concept of creatures fulfilling the wondrous outcomes of their potentials in accord with the will of the Creator.

If we gaze at the world in a spirit of reverence, we eventually understand that natural occurrences everywhere seem to exhibit a kind of obedience. Everything happens in response to nature searching for its realizations, not only as a whole but also in its countless components—even in its very smallest entities, the sub-atomic particles that work together to form the atoms. Everything occurs as if matter-energy and space-time were leaning toward fulfillment, almost as if trying to make their yearnings come true. And we who befriend atoms should bear in mind that the so-called "rules" of chemical reactions lend themselves to a more compassionate point of view, one where atoms, including those in our own bodies, can be understood as harboring yearnings on levels proper to themselves.

Our science textbooks, then, can be viewed as storybook accounts of creation seeking fulfillment in superbly consistent ways. The storybooks tell of things continuously role-playing throughout the ages as if having been programmed in the beginning to exhibit an endless-yet-not-monotonous, but rather exciting, sameness in their behavior, impassively awaiting change as time moves forward. It is as if things become what they are, and *are* as we find them, all of this happening along the avenues of instinctual tendencies at work within their atoms. Thus, we can say that the laws of chemistry and the periodic table of the elements express the "spirit" of atoms . . . not like the human spirit, certainly, but a spirit of a kind nevertheless in the sense of their consistent behavior revealing a fundamental readiness to act in persistent and committed ways. One can call it a spirit of readiness, if one wishes. All of this seems to reveal a sort of obedience that is perpetually at work within them.

Yet, lest we forget, the "rules" of chemical interaction are not regulatory axioms like the "do's and don'ts" governing human conduct. Things do not behave as they do out of an awareness on their part that they must follow rules. Whether alone or together, things mysteriously behave in superbly consistent ways, often as if listening to directives. How mysterious, all of this!

Physical laws are after-the-fact descriptive statements that we formulate to summarize the consistent modes of behavior exhibited in the material makeup of this world. Those laws say almost nothing about what these things might be tuning in to or listening to, or about what it might be that is enfolding them and bringing them to act as they do. And to declare, as British physicist Paul W. Atkins did, that things behave as they do "because they can behave in no other way" neither answers the question nor diminishes the mystery in the least.

Most persons lose their sense of awe regarding the ordinary ways in which things behave. A leaf, for example, becomes detached from a tree and falls toward the earth. Were I witnessing this for the very first time, I would certainly be surprised and charmed. But on seeing thousands of leaves falling to the earth across a lengthy period of time, I easily lose my sense of surprise and amazement. How easy it is for us to lose our fascinations when experiencing an abundance of what is wondrous! Viewing with appreciation the so-called insignificant things, such as an old discarded shoe or a not-worth-mentioning scrap of paper, often helps us to overcome our insensitivities. It is an excellent practice to habitually notice with appreciation the very simple things that people normally ignore.

To those with both a scientific and a religious bent, creation expresses something of ongoing presence in endless ways. In this sense, both the animate and inanimate can be viewed as expressive of that presence. How can we as Christians say this without fear of error?

To Christians who believe that Christ is Truth-expression in its fullest, there is strong justification for recognizing something of him in all creatures, living and non-living. Christians view creation as a boundless arena of truth-encounters with endless God-expressions all around us. We believe that the Holy Spirit aids us in understanding the world as truth-revelations coming to us, one after another in endless succession throughout our lives. By way of the truths that we recognize in limitless ways through different shapes, sizes, forms, colors, densities, odors, patterns, temperatures, sounds, rarities, and textures throughout the realm of "world," we grow in our understanding of the all-inclusive and universal Spirit of Truth toward which all of these things point simply by being what they are.

God is present to us not only in expected and intuitive ways, but also in convoluted, inverted, opposing, ridiculous, and unexpected ways. Truth in this world comes not only through what we find to be ordinary, pretty, exotic, fascinating, or "cool." It also comes by way of clumsy-looking giraffes, grotesque-looking deep-sea creatures, and donkeys that seem so ugly as to appear rather cute. Through them the Lord reveals to us, from diverse perspectives, something on the far side in the endless mystery of "I am." On seeing odd-looking creatures, then, we might well respond to the Creator by exclaiming, "Is *that* the way You expect us to derive our understanding of who You are? What strange creatures we meet! Tell us, what does each of them mean in terms of yourself and your ways?"

Finally, we might ask: Since humans are included in the concept of "world," what about human behavior and misbehavior? Can they be validly regarded as characteristic of the overall cosmic behavior? And, further: To what extent, if any, does the cosmos as a whole depend on humans for its inherent worth? This is an interesting question, and worthy of much reflection.

Future discoveries in science, no matter how radical, cannot nullify the recognition of a Presence in which all creatures exist as if rested or nested in it. Scientists will never discover a truth showing that there is no God, for God is Truth Itself in its fullness. The recognition of Truth as active and all-encompassing throughout the cosmos is the most fundamental perception of humans and grounds all honest judgments having to do with what is real.

For Discussion

1. It is well known that queen bees, if put together in a hive, will fight each other to the death until only one queen remains. Do you regard this as an aberration in nature that is "apart from the Lord"?

2. Were there no rational creatures such as humans to delight in the world, would the world still be "of wonder"?

CHAPTER 24

Evolution and Ascending Christology

"Man represents the most synthesized state
under which the stuff of the universe is available to us. . . .
To decipher man is essentially to try to find out how the world was made
and how it ought to go on making itself [under God]."
—Pierre Teilhard de Chardin

A s our searches for meaning proceed beyond the physical and extend into the metaphysical world, we discover a trend on the
part of reflective persons to view the supernatural and the natural as one. The metaphysical is sometimes seen as a furtherance or
culmination of the natural. With that in mind, let us address from the
standpoint of evolution Christ's coming into the world. Let us try to do
so as persons who are both science-minded and Christian. Let us consider both the world's behavior and the Lord's Spirit finding expression
in the stuff of the world, which, of course, includes our bodies that harbor our spirits.

Throughout the Christian tradition, Jesus the Messiah has usually
been viewed as having come down from heaven. Theologians refer to
this viewpoint as "descending Christology." On the other hand, a more
recent "ascending Christology" sees the Messiah like a second Adam, rising into the human family like ourselves, by way of evolution, and moving forward to ultimately draw all things unto himself (see John 12:32).[1]

Some theologians today are allowing for the possibility that evolution began with God placing a potency, vigor, or vitality into the makeup of original creation. And, thus, by way of the *ordinary* behavior of
matter-energy operating in space-time, Divinity would one day emerge as
a human person in this world. It could have happened in countless ways,
at an opportune or "appointed" time and at a specific place such as Bethlehem of Judea, in what is often referred to as "the fullness of time." This
view is wholly compatible with Christian faith, for Christians believe

that God came to walk the earth in the person of Christ, to discourse with people as a human like ourselves except for sin.

Such a profound theology that includes evolution bears significant implications for science. It would suggest that some of the ways in which God emerged into this world in the person of Christ are in fact being addressed, without explicit mention of God, in our textbooks of biology, physiology, chemistry, physics, astronomy, and geophysics. This idea might seem shocking to some, but it need not be. For the possibility of God emerging from within creation through evolution contradicts neither the belief that God is ultimately the reason for everything existing nor that Jesus is the Son of God, the Christ.

When, as Christians, we reflect on God's acting through evolution, ideas about matter, energy, space, and time are transformed. We are drawn into highly sensitive areas as we visualize the stuff of the world in the context of Christ's coming into the world by way of the world's ordinary behavior.

Pierre Teilhard de Chardin insisted that the concept of matter apart from spirit has no meaning. He described matter and spirit as part of the same evolutionary movement, but with spirit always leading matter:

> Matter is not the stable foundation of the world. Rather, it is spirit which holds the primacy. Everything holds together from above, not from below. For the man who knows how to see correctly, the analysis of matter reveals the priority, the primacy of spirit.[2]

Understanding God as somehow implanted in creation from its beginnings, Teilhard pictured love as the energy of unification, and thus identified God, who is Love in its fullness, as the underlying driving force of evolution, from which the Creator in the person of Christ would arise one day to lead the world toward final transcendence.

Theologian Karl Rahner considered evolution an open question. While avoiding actual endorsement of evolution, he proceeded to sketch a system of theology that would allow for Teilhard's outlook. His thinking is of value to persons who view the human race as evolved from the God-given potencies of matter. Rahner defined the human as

> the existent through whom the basic tendency of matter to know itself in spirit through self-transcendence reaches a definitive breakthrough.[3]

From the viewpoint of science, Teilhard's profound definition allows for innate tendencies toward higher activities residing in the basic particles of matter. Moreover, it stands as a fascinating example of how Christian theology, while remaining faithful to tradition, undergoes an extension (if not an evolution) of its own, holding itself open to speculative new thinking that does not clash with basic Christian beliefs.

Rahner's definition conjures up images of atoms everywhere striving to move beyond their present state, reaching always toward their zenith. Theirs is an activity of hope on a rudimentary level arising from the tendency of the substances of the cosmos to reach beyond themselves. Because we ourselves engage in that activity of hope, his definition includes all of us along with our surroundings.

Rahner goes on to say that it would be unchristian to insist that matter and spirit merely exist alongside each other. He further states that Christian theology assumes that spirit and matter have more in common than not in common.[4] And, as far as we presently know, it is especially in humans that spirit and matter are personally experienced as coalesced, working together as one.

Many persons have come to see and believe that our present experience of rational participation in the cosmos flows from an incredibly wonderful, spectacular, and unrelenting excursion through time. It seems to be the outgrowth of a developmental voyage reaching back billions of years, with further culminations yet to come. We can imagine that human self-awareness originated with reflective "I–it" encounters between ourselves and things around us, and that this extended itself into the conscious "I–thou" relationships arising between ourselves and others, with compassion entering the picture.

Human life is not as unchanging as we might imagine. In the overall orchestration of matter and spirit, it possesses a movement all its own. Teilhard put it this way:

> Man is not the center of the universe as once we thought in our simplicity, but something much more wonderful—the arrow pointing the way toward the final unification of the world in terms of life. Man alone constitutes the last-born, the freshest, the most complicated, the most subtle of all layers of life.[5]

And elsewhere he writes:

> Laboriously, through and thanks to the activity of mankind, the new earth is being formed and purified and is taking on definition and clarity.[6]

An exciting aspect of all this is the phenomenon of innovation. When viewed from a broad perspective, innovations are intelligible in terms of the cosmos acquiring never-before-actualized entities through us, its self-reflective cerebral components. Through each of our inventions, the cosmos takes a forward leap toward self-advancement. Through each breakthrough, something of the "I am" of the Lord enters the world in a never-before-realized way, and it is truly proper for us to regard ourselves as appointed, chosen, or assigned by the Lord to bring about such marvels. (And, lest we forget, this applies also to innovations on the part of small children at play as well as to the innovations of adults.) We are the brains of the cosmos, the highest known outcomes of its evolutionary process of complexification, its "cerebralized" constituents who enable it to indulge in knowledge of itself.

Discoveries as well as innovations, then, are evolutionary mileposts. They escort us into a new awareness of the divine presence in our cosmic odyssey, a journey in which the Lord is both benefactor and beneficiary.

Beneficiary? How can that be? In what sense can we say that the Lord is a beneficiary of our cosmic journey?

Given the fact that God willed to become one of us, then (from the standpoint of ascending Christology) we can reasonably say, at least poetically, that God "took advantage of evolution." Evolution can thus be viewed as a crescendo of a sort in the orchestration of God arriving at last as a human in our world, eventually making an appearance as a totally unique person born of a Jewish maiden. Here is the very heart of the sacred dimension of evolution! In this view, *the ways of the cosmos are the avenues* leading toward the emergence of the Creator in creation, the event known as the Incarnation.

If we think of evolution from a Christian perspective, we see that we are called to engage the world in God's name. When engaging its substance and circumstance, as we do at every turn, when navigating its settings and surroundings, we do so with the belief that Christ is relevant everywhere and in every consideration. And so, we recognize ourselves as called to reshape the world with him in mind. We do this not always by action, but sometimes by way of restraint.

There is very much more to say about matter and spirit, good and evil, time and eternity, and other such phenomena viewed through the notion of evolution. An unusual and scholarly treatment of these subjects—a sort of synthesis of physics, biology, and mathematics with philosophy and theology— can be found in a book entitled *Cosmos* written by my friend, Richard J. Pendergast, S.J. In his book, Pendergast speculates on evil as stemming from the very makeup of matter, even before the human race began. He focuses on what he calls "the fall of the universe," writing:

> Our present world is the result of an evolutionary process which
> is probably ten billion years long and which went awry from the

very beginning. As a result, physical entities on every level are partially isolated from one another and pursue their own individual good regardless of the effect it has on others.

... For us, the Fall must not be merely or "only" the fall of our own biological race but, rather, the fall of the entire universe. The event which twisted the world from the path God had set for it must have happened long before Earth existed, in all probability before the evolution of the elementary particles some ten billion years ago.

... Sin and death are written into our very genes in such wise that a sinless human life was from the beginning a miracle of heroism which could not be expected from a large group.[7]

His interesting perspectives give rise to further speculations. How might creation have gone wrong as viewed from the standpoint of science? In what sense might we correctly say that the coming of Christ rectified the "sin and death written into our very genes"? These, indeed, are strange questions that one can only pose and leave unanswered. At least for now!

But, after all, does not the cosmos itself at times seem to exhibit a strangeness of a sort? Often its catastrophes instigate a new life by the simple rearrangement of things, strange to say! Often its eruptions and devastations bear the seeds of fresh beginnings with outcomes that are more consoling than before ... a kind of inherent resurrection, we might poetically exclaim.

Ultimately, in blending the Christian's and the evolutionist's points of view, we are saying in effect that God, through Christ, is *personally* involved in our ever-changing world. This means that the good news of salvation that Christians acclaim is more deeply rooted in the substance of the cosmos than most people realize, for the world, possessing inherent weaknesses, has nevertheless sustained an unimaginably lengthy chain of events that eventually led to the appearance of God as a human. When all this is seen as arising out of the evolutionary activity of material substance understood as touched from its beginnings by God, then the phenomenon of "world" itself must be applauded as an entity of God-emergence and perhaps even redefined.

Our paradigms have shifted. Despite all its troubles, including our sufferings and death-diminishments, the world is recognizable as a majestic herald of God's presence. Not only is it saved from the "shipwreck of non-existence" as Chesterton put it, but also crowned with high meaning. God and world are properly understood as bonded together in

Christ. The concept "world" now includes Christ, and the glorified body of Christ is the highest derived entity from the substance of our cosmos.

Christians who view the world from the standpoint of ascending Christology see the coming of Jesus as a colossal breakthrough, a finalization of a celestial orchestration initiated eons before his arrival. Thinking in terms of the ordinary course of events happening as they do, Christians might sense wondrous outgrowths of convergence occurring in myriad ways. Reflecting on this from the standpoint of present-day cosmology, they would sense that from the initial fireball at the beginning of time, Jesus the Way was on his way through an upward tendency implanted in the ways of creation simply being itself.

Yes, science enters the picture here, for scientists play the foremost role in detailing creation's basic ways. But additional steps that take us beyond science are called for, discernment in particular. In the minds of Christians who believe in evolution, God through Christ is discerned as having been active, even immersed in both the inner and outer ways of matter, energetically at work in space along the course of time. And the lengthy process of evolutionary development from its beginnings can be understood as an advent of a sort, as a prelude to the coming of Christ. Since his coming, the definition of everything has been altered. Things everywhere are endowed now with a new dignity.

When thinking like this, it is well to keep in mind a certain philosophical principle, an adage that says:

> The more remote (or far-removed) a cause that brings about a truly wondrous effect, the more wondrous that cause must be.

With this in mind, one has very good reason for visualizing God creating the cosmos in what might seem to us far-fetched ways. Who among us can truly discern the ways of the Lord who is the Way? Nevertheless, it's delightful to try.

Notes

1. In using the words "ascending" and "descending" relative to Christ's origin, we are in no way referring to the concepts of heavenly origin in contrast to evil origin.

2. Quoted by Donald P. Gray in *The One and the Many* (Herder and Herder, 1969), 43.

3. Karl Rahner, S.J., *Foundations of Christian Faith*, trans. William V. Dacha, S.J. Seabury Press (1978), 181.

4. Ibid., 182.

5. Pierre Teilhard de Chardin, S.J., *The Phenomenon of Man*, Harper & Brothers (1959), 223.

6. Pierre Teilhard de Chardin, S.J., *Hymn of the Universe*, Harper & Row (1961), 93.

7. Richard J. Pendergast, S.J., *Cosmos*, Fordham University Press (1973). Quotations are from pp. 171, 168, and 170, respectively.

For Discussion

1. Does the idea of the Christ coming into this world along the route of evolution bother you?

2. Can one be certain that it did not occur along that route?

3. If it actually did occur that way, would this have made any difference in the validity of the teachings of Jesus?

CHAPTER 25

Navigating the World
of Transparencies

"While most people reserve their attention for grandiose spectacles—
the only ones worthy of interest, in their opinions—
the noblest minds have been bent over ridiculous bits of reality."
—Charles Damien Boulogne

S cience has had its share of noble minds bent over seemingly trivial things. Englishman C. V. Boys, for example, spent much of his life studying the behavior of soap bubbles and the forces within them that give rise to their shapes.

In a world where many look but few really see, the ability to visualize the "withins" of things is a very great gift. The fields of chemistry and nuclear physics are based on visions of the withins of matter. Scientists understand differences in materials in terms of the different kinds of particles that constitute them. They visualize certain physical and chemical properties in terms of the different kinds of bondings between atoms. Rising temperatures are viewed in terms of molecules speeding up; heat transfer by way of conduction is viewed from the standpoint of molecular agitations being passed along from molecule to molecule; microwave cooking is understood in terms of hydrocarbon molecules resonating; rust is visualized as iron and oxygen atoms being chemically united, producing molecules that readily reflect red and orange light to which we respond by exclaiming, "Hey, it looks rusty!"

Scientists understand the red/orange colors of sunsets as leftovers after the atoms in the atmosphere have scattered the blue component out of the white light from the sun. In terms of atoms absorbing and re-emitting particles of energy called photons, scientists know why light travels slower through glass and faster through air. And they picture the inner workings of transistors in terms of flowing "electron holes" (a migration of the *absence* of electrons). Strange imageries, indeed!

151

A further wonder of science is the recognition of structure in apparently empty space, not unlike what mystics do when they recognize God in situations where others look and sense nothing unusual.

My physics students sometimes performed an experiment in which they took voltage measurements at various spots in the empty space between two objects, such as a metal ball hanging above an oppositely charged metal bucket. Their data, when depicted on a graph, revealed a series of so-called "equipotential surfaces" that showed the space between bucket and ball to be electrically structured.

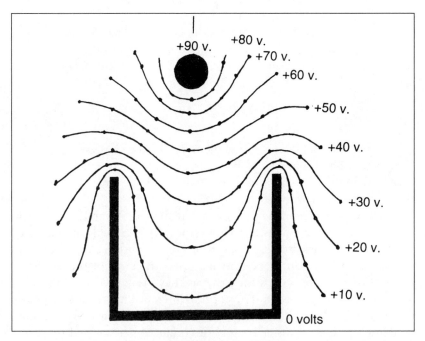

CAN EMPTY SPACE HAVE STRUCTURE?

This sketch drawn by students following a laboratory experiment shows the electrical equipotential surfaces they found to exist between a metal bucket and a metal ball hung above the bucket. The drawing reveals that what at first appeared to be empty space between bucket and ball did, in fact, contain a highly structured electrical field that was made visible by the experiment. The ball was maintained at +90 volts whereas the bucket was maintained at 0 volts.

When pondering the invisibles concealed behind the visibles, a new kind of language could be used. In the above example, one could state that a structured arrangement was hidden in the "implicate order." A question then arises: "What are the implicates that lie hidden behind the countless explicates of our world? Does this somehow bring us to include the spiritual world with the implicates?"

Scientists are familiar with the process of discovering invisible structures in empty space. Yet, beyond their visualizations of the unseens, there lies another and more ethereal perception of the material world that seems to elude even the minds of many scientists. It is a perception that Gerard Manley Hopkins termed "inscape," a word signifying a spiritual movement toward the essence of things. James Finn Cotter defines inscape as follows:

> Inscape is simply the pattern we all look for (or should) when we let the eye catch, or be caught by, beauty as it happens in a stone or a thunderstorm. Forms await there as soon as we become aware that matter is more than mere mashed potatoes...[1]

Inscape leads us into the experience of "transparency," of seeing through creatures to the Creator. Emilie Griffin writes of this transparency:

> Sometimes this experience of transparency is an experience of particularity. This one moment that I am now experiencing, this rainstorm in which I am being soaked and drenched, this feeling of mushiness in my shoes, this joke, these tears, this cat-ness of a cat jumping up onto a windowsill, this ability to know that the air is damp and that the leaves of the caladium plant, deep red at the center, green at the edge, are shining with wetness, through the garden window, just there, that very ability to experience the *this*-ness of things, is a glimpse of the real...
>
> A leaf, the root of a tree, the dime-ness of a coin glinting in the palm of my hand...this very one, no other! *This*-ness floods us with conclusiveness. Instant on instant, one following another, is Yahweh saying, "I am!"[2]

Viewing transparency from the direction of science, we might reflect on the attractiveness of gold, for example. At first we see only a particular piece of gold—shiny, yellow-orange, and smooth. Assuming we are open to wonder, we then move beyond the surface attractiveness of gold and appreciate it in both scientific and poetic ways for what lies "within," visualizing the 79 electrons, 79 protons and 118 neutrons of its atoms somewhat like parents gratefully discovering the ten toes, ten fingers, and two marvelous eyes of their newborn infant. The encounter is one of gratitude just for being in on the mystery.

Scientists speak confidently of unseens within unseens—of protons enclosed within atoms. Theirs is a shepherd-like engagement with the invisibles of this world. They go out into the wilderness of substance in search of its not-yet-discovered "lost sheep," rescuing the unseens from nullity, bringing them forward into the light of recognition. By doing this, they acknowledge the invisibles to be worth their time and worthy of appreciation.

In our scientific visualizations of the world, we make use of models that help us to picture the invisibles in terms of what we know as visible. Throughout the early 1900s, for example, physicists likened atoms to very tiny solar systems, with electrons orbiting around nuclei. Models, too, are used in the world of religion, except that in religion they are referred to as *myths*. Myths are employed in religion not because their subjects are untrue, but because those subjects are so profoundly true that we must couch them in stories and devise some sort of imagery to make their essence more intelligible.

As mentioned earlier, Johannes Kepler was entranced by the musical strains he imagined hearing in the orbiting of the planets. Today we regard this with amusement, for we certainly do not expect ever to hear real music emanating from planets undergoing orbital motions within the vacuum of outer space. Considerations such as this, however, bring up the notion of human imagination—in particular, our ability to gaze at things and fantasize far beyond what we actually see with our eyes. Humans have the unique ability to fantasize, to visualize the world in dimensions far beyond the obvious. Thus, human imagination ought to be of high interest to persons who think of nature in all-inclusive ways.

There is so much to say about the endless ways our world presents itself to us as we struggle toward visualizing it in its totality while seeing it only narrowly by way of its parts. Here are a few basic considerations.

Because so many of our observations of the world are made by means of light, our philosophizing about the cosmos might logically begin with the behavior of light as a prime consideration. What is light?

Light consists of countless tiny particles of radiant energy known as *photons* that enter our eyes, for example, from an object such as a sailboat. Trillions of photons each second from all parts of the sailboat must enter our eyes in order for us to view the vessel as a whole. But we do not perceive the vessel as discontinuous or grainy in appearance, for our visual nerve system presents us with a smooth image formed by the blending together or "averaging out" of information delivered by the individual photons.

While it is a worthy consideration to know the world by way of light coming to us from matter, and to reflect on meanings that arise by way of our perceptions of matter, an entirely different consideration also arises. What about the *lack* of matter? Can the *absence* of matter have transcendent meaning? This, of course, brings us to the consideration of space.

In real life, the concept of void or empty space is understood always in contrast to matter. Things that border or outline empty space come to be understood as describing the shape of space. Absence requires presence for its meaning, and so we learn at an early age to understand emptiness in terms of things not being where they conceivably might be. Imagination is especially needed for a good understanding of space.

Theoretical physicists have long engaged in fascinating visualizations relating to space. Beginning with Einstein, they have spoken of space as "curved" by the presence of matter. They have even gone so far as to speak in positive ways of things in terms of their non-existence. For example, a physicist might refer to electrons as "holes" in a universe of non-electron space.

Thinking for the moment in terms of creation, can it be validly stated that the voids between things are creations, too? People today speak of "outer space" with a remarkable familiarity. They imply, at least in certain instances, that absence of matter is part of what enables them to have freedom—at least freedom of movement. We seem to recognize spatial voids as giving depth to creation, and to visualize "wide open spaces" as offering us liberty. Let us illustrate with analogies from the world of music.

When we listen to melodies, not only do the actual notes that we hear have meaning, but also the intervals of silence separating the notes become a part of the overall composition. They hold rich meanings for listeners who recognize them as strategic pauses in the overall flow of melodies.

In their compositions, composers delve deeply into the world of tonal relationship. They create their music in conformity with the succession of moods they want to express. While composing, they remain always aware of the consonance and dissonance of combined pitches, the need for repetition, the loudness, and the tempo within an endless world of tonal qualities, note combinations, and pauses between notes. By shaping these, they can communicate to listeners a succession of moods ranging from rage and depression to joy and peace.

With regard to the fascinations of composing music, Suzanne Langer reflects:

> In music the fundamental movement has the power of shaping the whole piece by a sort of implicit logic that all conscious artistry serves to make explicit. The relentless strain on the musician's faculties comes chiefly from the wealth of possibilities that lie in such a matrix and cannot all be realized, so that every choice is also a sacrifice.[3]

Many contemporary thinkers in the field of creation spirituality are conceptualizing the cosmos aesthetically as a kind of "grand symphony" in which people play a weighty role. Accordingly, our acts of driving an automobile, opening a can of tomatoes, raking leaves, or taking our shoes to a repair shop can be thought of as melodic renditions against the background orchestration we call "world." This poetic view of the cosmos conjures up innumerable imagined musical notes of countless pitches and durations, individually and collectively working together in three or more dimensions of space-time. Every atom, molecule, and photon throughout the whole of creation is envisioned as uniquely contributing its song. Every leaf, star, insect, blade of grass, windowpane, broken twig, sparrow, bulldozer, railroad crossing, discarded tin can, tree stump, and, of course, human person helps to dramatize the glorious rendition. One might ask what their music would sound like if orchestrated together all at once. The outcome, of course, would depend on how fertile an imagination one might have.

An added delight in this imaginary view of the world is that it models the cosmos after something already within the cosmos—namely, the sounds of music that linger in the human mind. To correlate the cosmos with beautiful music is to view it with reverence by way of its observable parts, each corresponding to individually emitted musical notes. It is particularly wonderful to do this while understanding the cosmos to be very much more than any of its parts! This is also analogous to the ways in which persons of faith visualize the Lord God through beauty in creation. We do so always with the understanding that God is infinitely more beautiful than the beautiful things we see in particular sectors of creation.

In contrast to the prevalence of false materialism in our world, new views of matter-energy and space-time—the constituents of the cosmos—are emerging. Many people are beginning to visualize the world as endowed with higher meanings by way of matter appearing in remarkable new forms. Pocket calculators, computers, television sets, cellular phones, and microwave ovens are examples of things viewed today with appreciation . . . especially by those who remember their struggles in the days before they had such magnificent God-expressions.

A fascinating article appeared recently in *Scientific American.*[4] "Is the Universe out of Tune?" explained how scientists "listen" to the music of the cosmos by monitoring the background radiation arriving from outer space. These arriving microwaves are analogous to those of sound waves, but are indicative of variations in temperatures throughout the universe rather than variations of audible sound. The article goes on to say that the actual frequencies of these waves are somewhat different or "off key" from what has long been theorized, and the question arises as to whether or not the universe is "out of tune."

In the world of music it is good that a composition does not consist of a single note sustained for a very long time. So it is in the world of the material. It is marvelous that the cosmos is not a single expansive glob. It is presented to us in parts that are separate and interactive, so that the spaces between earth and moon, between the tips of snowflakes, between points of origin and places of destination, between salt and the food that salt is meant to flavor, between hamburgers and the teeth that we use to bite into them, all harbor deep significance. Not only matter itself, but also the voids between its bits and parts have latent meanings. It is as if the voids have enjoyed a secondary creation that is uniquely their own and await humans to respond to their magnificence by crowning them with recognition.

A novel example of creation being poetically affirmed in terms of its many parts is this expression by G. K. Chesterton:

> It is the instinct of Christianity to be glad that God has broken the universe into little pieces, because they are living pieces. It is her instinct to say "little children love one another" rather than to tell one large [cosmic] person to love himself.[5]

Notes

1. James Finn Cotter, *America Magazine* (January 21, 1984), 32. Cotter was a professor at Mount Saint Mary College, Newburgh, NY.

2. Emilie Griffin, *Clinging: The Experience of Prayer*, Harper and Row (1984), 31.

3. Suzanne K. Langer, *Feeling and Form*, Charles Scribner's Sons (1953), 122.

4. Glenn Starkman and Dominik Schwarz, *Scientific American* (August 2005), 48.

5. G. K. Chesterton, *Orthodoxy*, Dodd, Mead, and Co. (1908), 132.

For Discussion

1. Can you cite an occasion in your life when you experienced a "transparency" or eloquency of a sort in relation to an object, a scene, or a person?

2. What are some of the ways in which you visualize the world as "an unchained melody, the work of an orchestrator"?

3. In your contemplations, do you ever visualize time itself as having structure?

CHAPTER 26

On the Worldliness of God

"I watched a hawk soar, a flower unfurl its petal,
a fish arc high out of the water . . . and Jesus was on my mind.
'All things were made by him,
and without him was made nothing that has been made.'"
—Jerome Ledoux

L ike it or not, we are deeply immersed in the currents of this passing
world. The world, however, can serve to enhance our spiritual perspectives. To those of us who listen with discernment, the world presents itself to us as expression of God.

The above three statements comprise an overview of this chapter. In
the following reflections, each of the three will be considered in turn.

First, we are deeply immersed in the currents of this passing world.

It is well known that monastic Christianity harbored a long-standing
tradition of fleeing the world for the sake of overcoming evil and gaining Heaven. Yet it cannot be denied that scientists, as persons caught
up in the intricacies of this world, are deeply immersed in the lure of
passing things. We devote our lives to observing, investigating, manipu-
lating, and understanding the inner and outer substance of this world.
Might it be, then, that scientists are lacking in spirituality because of
this?

Much of the Christian distrust for the material world, we are told, is
a result of certain writings of St. Augustine (AD 354–430) who, prior to
his conversion to Christianity, was a Manichaean. Manichaeans held
that matter is evil whereas spirit is good, that the human soul seeks to escape from the kingdom of darkness that is the body. Thus, the di-
chotomy between matter and spirit to which Augustine adhered has
come down to us, and its influence is still very strong.

Many centuries after Augustine, a classic writing of Christian spirituality entitled *The Imitation of Christ* was written. Attributed to a monk named Thomas à Kempis (d. 1471), it begins:

> Turn with your whole heart to God and abandon this wretched
> world, and your soul will find peace. Learn to despise that which
> is without, and give yourself to that which is within...

It would seem from this that one is being advised to abandon the very ship on which one is a passenger while crossing the ocean of life. Misunderstandings persist even today over the various uses and meanings of the word "world."

It is unfortunate that many Christians imagine that immersion in this world is synonymous with selling their soul to evil. According to Teilhard, however, the reason why many abandon themselves to false materialism is that they have immersed themselves self-centeredly into the currents of the world.

Inasmuch as our bodies are composed of matter, we do engage in a proper materialism when caring for ourselves. So let us make a distinction here. Materialism becomes false only when it entails mindless surrender to the lure of passing things with little consideration of higher values. When appreciation, moderation, sharing, and self-restraint are not embraced in relation to things, a wrongful materialism results. As Chesterton once suggested, the problem with the worldly is that they do not understand even the world.

The ways of visualizing material are many, and the levels of understanding with regard to particular things are numerous. Consider, for example, the following:

A young soldier who happens to be a Catholic is encamped with his battalion along a ravine on a Sunday morning. He decides to attend the worship celebration that in his religion is known as "Holy Mass." But this particular Mass differs from others that he has attended because it takes place outdoors.

On this particular Sunday, the battalion chaplain arrives at the encampment and, on spotting a large boulder, decides to use it as his altar. He proceeds to conduct the ceremony in which, according to Catholic belief, bread and wine are changed into the body and blood of Christ, who is worshiped as the one and only Living God. An hour later the chaplain departs.

Several days later, the soldier returns to that setting and, in a contemplative mood, gazes at the large rock. He sees it as most others see it, as simply a sizeable rock. But, additionally, he views it as a sort of shrine, a personal sacred place of his, a location of high significance where the Lord recently stepped into his life in a special way.

It happens that this soldier also holds a degree in geology. Through his understanding of the earth's materials, he proceeds to visualize the rock according to its internal makeup—its crystals of quartz, its microcline and silicon molded together by their inner forces of molecular bondings. Having studied those things in the classroom and laboratory, he is much more deeply immersed in that material realm than are his fellow soldiers. He is clearly capable of viewing the rock in ways that they cannot. His science, along with his religious beliefs, have led him to view that small segment of the earth in unusually profound ways.

Karl Rahner insisted that Christianity makes us "correctly more materialistic"[1] than people in the pre-Christian Greek age. Aware that God through Christ has become one of us in this world, Christians have valid reasons for understanding material things as related to *God's worldliness* through Christ. And correspondingly, we can understand Christ from the standpoint of God having laid hold of material substance at the level of humans precisely, as Rahner says, "at that [evolutionary] point of unity at which matter becomes conscious of itself," where humans emerge and "spirit possesses its own essential being in the objectification of matter..."

There is a well-known Christian saying that the Holy Spirit sent by Christ has "renewed the face of the earth." In this writer's opinion, many who use this expression do not understand the richness of its meaning. At the very moment when God took on a human nature through Christ, God also became *intrinsically* united to matter through Christ. Thenceforth humans who believe in this mystery have reason to visualize *all material everywhere* as aglow in relationship with Christ. Concern for Christ becomes our driving force for reinterpreting everything everywhere. Christ is our reason for speaking of material as God's own by way of the Incarnation, the peak event in the coming-to-be of God's personal enmeshment in matter.

Once we understand God as having become personally immersed in the currents of this world through Christ, we need not fear that immersion in worldly currents will result in our spiritual ruin. After all, the material of this world now harbors the seeds of salvation. In a mature Christian consciousness, anything anywhere can now bring Christ to mind. And our world becomes open to rediscovery in terms of this relationship to God. Teilhard wrote:

> Our faith imposes on us the right and obligation to throw ourselves into the things of the earth ... to prolong the perspectives of [our] endeavors to infinity. Anyone who devotes himself to human duty

according to the Christian formula, though outwardly he may seem to be immersed in the concerns of the earth, is in fact, down to the depths of his being a [person] of great detachment.[2]

One of many ways in which things serve as witness to God's goodness is through their availability. Thus, the merchants who make the world's goods available can claim a rightful dignity in doing so. And when we recognize their goods as living up to realistic expectations, we reveal something of the genuine hope that we place in the Lord through the things of this world.

Similar statements could be made in regard to countless other professions, assuming that our underlying concern is aimed toward the higher good. Machinists, for example, engage in fashioning materials into more desirable forms corresponding to the particular needs of people. Economists engage in concerns for fairness in the distribution of goods. Repair personnel work to revitalize things on which time and use have taken their toll. Gas station attendants make it possible for long-distance travel to occur, allowing people to experience a wider range of creation.

Clearly, the delights we experience through the material—the shoes we wear, the eyeglasses through which we gaze, the steering wheels we use to control our cars—rest not so much in the quantity of things possessed as in their charisma. We often delight even in the charisma of things that we do not personally own, and also in imagined things, such as a home in the planning stage, that do not yet exist. The recognized loveliness of things—for example, a well-located and properly installed electric wall switch—is a kind of revelation of God's goodness, and therefore of God, within our psyche. To the extent that we are graciously receptive to the innuendoes of earth, air, sky, and all other things, we enable them to deliver revelations of God's concern for us. It all depends on our openness to higher meanings in our search for that which is wondrous.

Artists exemplify people addressing material with reverence, utilizing certain substances to portray the world in wondrous ways. Typically, they are persons who have been awakened, as was Joan Puls,[3] to the fact that "every bush is burning."

Second, the world, however, serves to enhance our spiritual perspectives.

Unfortunately, precious little is ever said about matter's lovely ability to enhance our spiritual perspectives. But there is much to be explored in this area. For example, why do we gaze at stars, ocean waves, and wheat fields and experience a sense of mystery? When doing so we allow them to lead us into new dimensions of awareness. This, of course, takes us back into the "inscapes" and "transparency" of things mentioned earlier.

I walk past a construction site and ask myself what might be the deeper meanings in the ordinary actions of carpenters who are hard at work on a half-completed house. A certain skinny man with long hair catches my eye. He pulls out a tape measure, determines the length of a board, cuts it, and nails it in place. I ask myself what he might be doing as viewed from a more spiritual perspective. What might be some of the inscapes of his actions as seen from the Christian point of view?

I remind myself that to measure an object's length is to enter into its world of dimension, to empathetically engage in its cosmic characteristic of extension. It is an activity of participation in the entitlement of things to their space. To cut a board to a specific length determined by oneself is to recognize its vulnerabilities, to lay hold of its possibilities, to determine its specific mode of servitude, to understand and to do something about how its potential will be expressed.

To fit two boards together before nailing them into place is to have their surfaces interact with each other in search of alikeness. Compatibility and complementarity are at issue. Concern over their proper fitting together is a kind of solicitude for harmony between them, ultimately with the future owners of the house in mind. The final nailing together of the boards is an act of endorsement of their harmony, a crowning of their accord.

Karl Rahner assured us that God and things are not simply two entities alongside each other. God is identifiable with "world." Yet there is a difference between God and world—a difference which, according to Rahner,[4] comes from God and is identifiable with God.

Rahner implies that it would be erroneous to think of things as merely symbolic of the Lord, as if nothing of the Lord is alive and active in them because they are only things. Just as we might hesitate to speak of an ambassador as merely symbolic of her nation, so we should also refrain from thinking of things as merely symbolic of the Lord, for Christ the Lord, in a very real and non-pantheistic sense, is understandable as God-become-world. In taking the world unto himself, he has embraced material in the settings of time, and has taken on the childlike characteristic of "getting into everything."

Again, questions arise: From a spiritual perspective, then, what is this entity that we call "world"? What are some of its emerging definitions that enter the consciousness of maturing Christians along the route of creation spirituality? The following are but a few. The cosmos is:

- That through which God communicates something of the invisible Way to us in endless visible ways

- A visible expression of invisible Love

- The place where our spirit and personhood are woven into the makeup of matter and energy at work in the realm of space and time

- That which outwardly expresses countless modes of being and thereby reveals something of the Lord, who is Being in its Fullness

- That which is a referent toward understanding something of the Lord's forgiveness as time and substance repeatedly offer us new beginnings

Viewing the world from a transcendent perspective, environmentalist Thomas Berry writes:

> ...It is clear that the universe as such is the primary religious reality, the primary sacred community, the primary revelation of the divine, the primary subject of incarnation, the primary unit of redemption, the primary referent in any discussion of reality or of value.[5]

It has long been said and often repeated that in leaving this world we cannot take it with us. Yet there are good and valid reasons to believe that, from the spiritual perspective, we shall indeed take this world with us into the next world, but in a different way than generally imagined. The flowers, the pine needles, the picket fences, the frying pans, the pianos, the rubber bands, the silicon chips, the tricycles and skates of our childhood—to people of faith, these things have an enduring beauty by way of the distinctive services they render in helping to mold our spirits during our life on earth. Those of us who are gifted with appreciation shall take with us the imageries of our best sweaters and worn-out shoes, our measuring cups, our musical instruments, our flashlights, our best-prepared foods, and even our barking dogs and long-suffering neighbors! We shall take them with us, not in the sense of physically toting them into the next life but, rather, in the broader sense of treasuring them endlessly for what they did for us in keeping awake our expectation of Goodness during our worldly pilgrimage. We shall possess them through their charisma, what they portrayed of the being of the Lord in concrete ways by way of their mystique being correctly discerned.

Within the innermost recesses of our being, we shall retain and harbor the things and situations of this world, including our apparent setbacks, understanding them as having formed the framework of that which kept alive our hope of seeing God face-to-face. For we shall have dimly seen something of divinity through each of them, including our sufferings, while on earth. Through their own specific charms, their beguilements, their dangers, their allurements properly understood, we shall possess them in new ways and crown them for the presence of God that we sensed through them.

Christians shall finally recognize that Christ, as God-become-world, was the highest expression of the physical world in which he and they once roamed. God will be understood as the fulfillment of everything that the world had been trying to tell us. We shall understand that once God personally joins the world, then the world's highest meanings necessarily inhere in God.

Third, to those who listen with discernment, the world presents itself as an accumulation of God-expressions.

What follows are a number of spiritual musings and intuitions on this subject. As an introduction, let us consider an insight proposed by the late R. Buckminster Fuller:

> God, to me, it seems, is a verb [and] not a noun, proper or improper.

One day long ago, a scripture professor of mine mentioned that the ancient Israelites interpreted creation's activities as direct personal expressions of Yahweh, their God. According to this professor, when speaking of water flowing in a river, or of inclement weather, an ancient Israelite would likely have said, "Yahweh is flowing" or "The Lord is raining!" In contrast, we today refer to those phenomena by saying, "The river is flowing," or "It is raining!"

In contrast to the ancient Israelites, we commonly avoid mention of a deity acting directly through the things that we perceive. But it need not be so, and perhaps *ought* not be so. Indeed, we can take definite steps toward moving beyond present-day secularized views of the world.

For example, borrowing from the ancient Israelites, an astronomer today might address a group of visitors in a spirit of wonder: "Come and gaze through this telescope! Look and see how abundantly the Lord is 'starring' out there tonight!" Or, again, someone might exclaim while walking through a park: "Isn't it magnificent how the Lord God is 'oak tree-ing' out here?"

Also, from this perspective we can imagine ourselves as walking along a street and gazing with interest at flowers and fences. Instead of exclaiming, "What beautiful roses! What odd-looking fences!" we could visualize those things on deeper levels of awareness and exclaim, "How beautifully the Lord is 'rose-ing' out here! In what odd ways does the Lord 'fence' into this world through the hands of people!"

What we are doing, of course, is taking nouns and transforming them into verbs, the actions of which are attributed directly to God. In this mode of awareness, creatures are pictured as immediate, direct, and active expressions of God.

There are endless delights in the experience of viewing creation in this manner. If we are to discover humor in spiritual escapades, this is one of the domains where it might be found. For example, we might ask what it means to construct a chimney or bake a cake. Perhaps in performing these tasks we are activating the Lord's "chimney-ing" and "cake-ing" into this world. Such thinking seems wholly compatible with Christian faith, for nothing in the whole of creation is aside from the Lord.

Let us recall once more the viewpoint of psychologist Carl Jung, who said that we know the world's things only insofar as we perceive their psychic images within ourselves. If this is the case, then a person of faith can reasonably ask a series of questions such as these: Might it be that our psychic images of the world are received, assimilated, and understood in the innermost depths of our souls as self-expressions of the Lord communicated to us through the medium of "world"? Can we validly regard subjects and objects throughout this world as verbs of God? Assuming that the answers to these two questions are yes, might we then have good reason to think of the human psyche itself as an interpreter directly translating "objects perceived" into "God-expressions recognized"? Might it be that in sensing people and things all around us we are in truth recognizing what is expressive of *ongoing accomplishment* on the part of God?

On a personal level, I have come to believe that, as God-become-world, Christ is at the heart of the Christian psyche as it translates "world observed" into "God-expression perceived." And I have often wondered if a maturing Christian might validly conclude that the personhood of Christ is the goal and fulfillment of knowledge itself. For I have come to believe that knowledge as a phenomenon totally rests in Christ who has identified himself as Truth Itself (John 14:6).

I ask myself, "Can I justifiably and profitably visualize the unlimited constituents of the cosmos in all their mind-boggling variety as direct expressions of the Lord God? Could each one be, in its own unique way, a revelation of God's 'I am'? And can I genuinely think in this manner without engaging in pantheism?" Whoever would answer these questions in the affirmative would certainly find this kind of mental engagement to be exciting. And, not surprisingly, such reflections also afford opportunities for humor…

We can ask ourselves: When gazing at crabs, worn-out shoes, or junked washing machines, cannot we reasonably and lovingly tease the

Lord about the ridiculous ways in which these creatures "speak the Lord" into the consciousness of people? Indeed, there is much to ponder regarding the very meaning of the term *ridiculous*. Recall once more (from chapter 5) how Chesterton once wrote that the rhinoceros looks "as if it does not exist."

Our richness of experience in visualizing things as direct expressions of God is beyond simple description. Every blade of grass becomes a particular expressing-into-this-world on the part of God who can be understood as repeatedly "grassing" into physical reality across lawns, parks, and pastures. Different varieties of grass, in the light of this understanding, would express the diverse ways in which the Lord "grass-expresses" into the world of human consciousness. For us, then, to classify the things of this world is to group them together according to perceived similarities and differences in God's ongoing self-expressing. How dynamic! What joy can be found in the ordinary things that we do when translating the meanings of our actions into visions such as these!

As applied to our perceptions of fellow humans, it would be fully proper for us to think of one another as God-expressions in the world of animal-rationality. We might understand the Lord as "Jane Doe-ing" into this world through a particular young lady named Jane Doe, with all her particular looks, talents, and peculiarities. And one of Jane Doe's cookies would be understood in the context of God "cookie-ing" into the world along the route of God's "Jane Doe-ing into this world." Also, we might say that the Lord has "Rachmaninoff-ed" into the world of music through Rachmaninoff, and that God then entered the consciousness of music lovers in uniquely beautiful Rachmaninoff ways. It's all so fantastic when we view our world with appreciative imagination—like living in a wonderworld!

Those attuned to visualizing things as divine assertions experience joy over the simple realization that we thrive in a world of people and things. They see deep and exciting meanings in nature's repetitions. When sunning themselves on a beach, they do not overlook little things—the grains of sand, for example. They delight in the silicon-God-expression of the individual grains and picture them in terms of their exemplifying specific instances through which Being in its magnificence is "grain-of-sand-ing" into this world in this particular place and time. They view beaches in terms of billions of sand-grain God-expressions clustered together in their similarities and differences. And if they think as scientists of religious faith might think, they see the beach not only as a place for having their usual fun with friends, but also as a place of wonder—a place where myriads of silicon grains form the enclosing boundaries for colossal numbers of water molecules imagined individually, then perceived as *drops* of water, then collectively as an *ocean* of staggering dimension. And they can pray from the depths of their hearts:

> Lord, might it be that in knowing particular things and people, we are knowing you, the Way and the Truth, in correspondence to the ways and truths of these things and people? Might it also be that what we know in this world as nouns are, in fact, verbs in your sight—actions on your part that blossom into being, each and every one in its own time and place? Can it be that every-thing in its own ineffable way—every lake, every mountain, every toothpaste tube, every empty milk carton, every violin bow, every elephant tusk, every cave, satellite, skillet, hosepipe nozzle, paper clip, speedometer, tent peg, flower pot, ocean freighter, beer bottle, discarded automobile tire, insect, animal and human person—is a firmly-whispered salient word-of-God spoken in the language of "world"? And is Christ, your Word, being spoken everywhere in endless ways at every moment through the language of the truths that each of these portrays?

Notes

1. Karl Rahner, S.J., *Foundations of Christian Faith*, trans. William V. Dych, S.J., Seabury Press (1978), 196.

2. Pierre Teilhard de Chardin, *The Divine Milieu*, Harper and Row (1960), 39 and 40.

3. Joan Puls, O.F.M., *Every Bush is Burning*, Twenty-Third Publications (1987).

4. Rahner, *Foundations of Christian Faith*, 63.

5. Anne Lonergan and Caroline Richards, eds., *Thomas Berry and the New Cosmology*, Twenty-Third Publications (1987), 37.

For Discussion

1. How often do you deliberately gaze at something ordinary and common with the intention of recognizing it in new and special ways?

2. Write out an ordinary experience of yours. Then write it out once more, translating it into the language of transcendence (as was done in the above example of the carpenter at work, page 162).

CHAPTER 27

Can Science Be Humanized?

*"If we see the meaning of something,
then that thing has in some way been changed."*
—David Bohm

Protagoras once said that man is the measure of all things. One might be tempted to ask whether our sciences, while bringing us untold benefits, might also be making us less than fully human.

Naturally, the concept of human person includes much more today than it did in the time of Protagoras. In response to inspirations, we meet the world and remake it in endless ways that differ very much from the ways of the ancient Greeks. Yet there persists something timeless in our re-makings of the world, something sprouting forth from our humanity that takes us into the realm of *beyonds*.

Professionals in the humanities—the scholars in such fields as literature, philosophy, and the fine arts—often insist that science has encouraged the idea of an absolute separation between mind and matter. Their complaints are numerous. Some insist that Kepler's formulation of the laws of planetary motion suggest falsely that the solar system is under our domain, that we are its masters inasmuch as we have figured out its behavior. Some also maintain that Isaac Newton, thinking of the world as "God's machine," has promoted deterministic imageries of creation. Such complaints certainly merit thoughtful consideration.

Christopher Bird has criticized what can be considered a certain arrogance on the part of many of the world's scientists. He complains that science itself has become religiously authoritative:

> It has what amounts to a priesthood, a hierarchy that decides for its laity—which means the rest of us—not only what is reasonable and therefore to be countenanced for study and funding, but what is taboo and therefore to be anathematized.[1]

On a lighter note, Michael J. Cohen[2] observes that because in nature the only constant is change, mathematical analysis of nature is never wholly accurate. Inasmuch as one tree is never exactly identical to another tree, it is not fully correct for one to say that "one plus one equals two" in regard to trees. The closest a person can come to the concept of "one" in nature is in terms of the whole—the whole earth, the whole cosmos. The holistic view is the most nearly correct.

What, then, are humanists recommending for reversing past mistakes? What steps need to be taken toward the "re-enchantment" of the world?

Faced with such a task, we might begin by asking whether humanists can be found within the science community. If so, what might *they* be saying about the inhumanities of science?

Certain modern scientists, few in number, have begun writing on this subject. Concerned with expressing their humanism along with their science, they search for ways to theorize their science with human harmony and social stability in mind.

An example of such a scientist was the late quantum physicist David Bohm (1917–1992), an exceptional intellectual who spoke openly about science in the overall context of meaning. In this chapter I shall be referring often to Bohm so as to familiarize the reader with his unusual and fascinating views.

Bohm insisted that the meaning of things constitutes their very being. He suggested that meaning is operative at different levels and insisted that everything is "an unbroken and undivided whole movement." He wrote:

> ...mind and matter are not separated substances. Rather, they are different aspects of one whole and unbroken movement. In this way, we are able to look on all aspects of existence as not divided from each other, and thus we can bring to an end the fragmentation implicit in the current attitude toward the atomic point of view, which leads us to divide everything from everything in a thorough-going way. [3]

Throughout the twentieth century scientists authored numerous works on the meaning of observation as they found themselves drawn ever more deeply into philosophical considerations that sometimes bordered the theological. Drawn ever closer to reflection on ultimates, those who wished to speak also as humanists addressed such questions as: What are the things that count as phenomena? Can we, as humans, validly understand ourselves, in simply knowing things, as thereby using them? Is putting things to use an essential consideration in our understanding of

QUANTUM PHYSICIST DAVID BOHM
(1917–1992)

Born in Pennsylvania, David Bohm was fascinated at an early age by the order and regularity of the universe. On graduating from college, he went on to receive his doctorate (1943) from Berkeley. He then performed research under Robert Oppenheimer and worked on problems connected with the Manhattan Project. After that he taught physics at Princeton University.

The interconnectedness of things impressed Bohm deeply, and led him to novel ways of looking at the universe. In time, he published his views on an "implicate order" existing throughout the space-time continuum of the universe, out of which apparent "explicate orders" arise.

His ongoing concern was to develop concepts and modes of thinking that relate with the experiences of people. His work often drew him into areas of speculative philosophy and, at times, touched somewhat on theology. This chapter gives the reader a few insights into the thinking of this great man.

things as having meaning? Is their "just being there" understandable as a service that things render to us?

When thinking in this manner, according to Bohm, we are led to ask: How much of science is *things*? How much of what we call physics is *thought?* How much of it lies outside of us? How much within? To what extent are our sciences fashioned by ourselves? To what extent are they products of our methodology? And to what extent are our methodologies forged by things? Unusual questions, but quite important!

Certain scientists dismiss such evaluative considerations as inappropriate. Perhaps they think that inclusion of the human person in their cosmic reflections would draw them into endless other questions, such as that of matching their science with the total realm of human experience. Indeed, in today's interpretation of what it means to say "world," this happens to be a kind of hot issue among certain scientists. Seemingly, most prefer not to discuss it. Yet the matter persists. The question of matching our sciences to our humanity won't go away.

Granted that fragmentation exists, how is one to mentally proceed beyond the impasse and come into the world of wholeness? How are we to think and form our sciences in ways that keep them in harmony with ourselves at our best? After all, we are creatures of nature who wish to properly discern what nature tells us about itself, which includes ourselves.

David Bohm conceded that the atomic point of view is divisive. Yet he insisted that it has a validity inasmuch as it provides insights into the various patterns of nature, such as those observed in crystal growth where atoms come together in orderly close-packed arrangements.

Bohm recognized that the ways in which we have traditionally looked at things exert tremendous influence on the ways in which we will look at them in the future. He therefore suggested that many of the old orders of thought in physics might cease to be relevant with the perception of new orders, new measurements, and new imageries. He recommended that scientists seek new approaches to understanding our world from the direction of aesthetic perception, that we seek to discover the artist within ourselves while addressing nature's behavior. Apparently, a continual discernment of nature as inclusive of ourselves would be needed to formulate new theories lest we repeatedly tack new insights onto old structures.

Bohm saw scientific fact and theory as different facets of a whole. He suggested that we be open to changes with regard to what is meant by *fact.* Facts, he implied, have been structured by our manner of interpreting physical reality in certain accustomed ways; we ought always to be aware that reality never fully conforms to the statements that we call "facts." In this regard, he wrote of scientific measurement:

> ...measure is an insight created by man. A reality that is beyond man and prior to him cannot depend on such insight. Indeed, the attempt to suppose that measure exists prior to man and independently of him leads...to the "objectification" of man's

insight, so that it becomes rigidified and unable to change, eventually bringing about fragmentation and general confusion.[4]

One of many ways in which we can rise above old orders of thought is to redefine, or *re-factate*, our understanding of science. But how can one re-factate what is already established as factual?

To a certain extent, we continually make and remake our facts in science inasmuch as we fit and interweave them into the framework of our previous thinking. On perceiving new facts, we assimilate them and extend them into our previously structured fields of meaning. We do, indeed, have a measure of control over what they will mean to us. For, in our normal ways of thinking, what we believe a thing *is* and what a thing *shall mean to us* are separate considerations.

Bohm maintained that we ought to receive new facts not as isolated data but rather in the context of wholeness. To better understand what he meant by this, let us consider, for example, the fact of the hydrogen bomb (this writer's example) from the standpoint of physics. This fact can be altered or extended by understanding it more holistically as also a political fact, a military fact, and a psychological fact over which people sometimes agonize. When working on the bomb, physicists engage also in the "beyonds" of the bomb's physics. If they are persons with a healthy outlook, they are called to recognize the implicit as well as the explicit aspects of data gathering. They are thereby called also to re-examine and re-factate the weapon and its delivery system in the context of wholeness. This would include concerns for justice, politics, the environment, fears of destruction, searches for good, and proper uses of the bomb, along with other factors affecting the human family.

It is important that our sciences not be so theorized as to end up dehumanizing us by hindering our psychological and spiritual well-being. Otherwise, our sciences stand as potential threats to us, the very creatures in whom the studies of nature rest, and to whom these studies ought to be matched.

Bohm uses the term "holomovement" when focusing on "undivided wholeness of flowing movement." He insists that the prevailing trend in modern physics goes against this undivided wholeness. What he denoted as holomovement centers on nothing less than that which is. Everything that exists is to be explained in terms derived from this holomovement. As we make discoveries in certain areas (or sub-totalities), we are not to regard them as having absolute and final validity. Rather, *the whole* must be kept in mind and never dismissed. This became particularly true with the discovery of nuclear fission.

Bohm believed that "a total order is contained in some implicit sense in each region of space and time." Each region embodies a total structure enfolded within it. He suggests visualizing the world in terms of an implicate, or enfolded, order that is beyond time and out of which time itself emerges. Its counterpart would be the explicate, or unfolded, order. From the world of television he gives examples of the implicate and explicate orders. Television waves traveling through space, he points out, can be thought of as enfolding the visual scene in the implicate order. The final displayed picture exemplifies the explicate or unfolded order that has been made explicit, poetically speaking, as if called from "out of the haze."

On reading Bohm's thoughts we may wonder: Might it be, then, that everything in creation can be validly recognized as having been, in a sense, called forth from out of the haze? Does not this understanding bear similarities to those expressed by the writer of Genesis, who visualized the Lord God as having called forth creation from "out of the void"?

When reflecting on the concept of the implicate in nature, we are drawn toward considering the ultimate in generative order. Such a consideration leads some scientists to wonder whether their intuitive ideas concerning an ultimate referent in nature correspond to the mystic's concept of God. There is much here to ponder.

What, if anything, do things draw upon in order to behave as they do? Are we to think of the universe functionally as a system closed unto itself without connections to a *beyond* or to an *outside influence*? And, again, in what sense would this outside influence really be "outside"?

I think the concept of nature being directed in an ongoing fashion by a higher being bothers many scientists. Some concede, however, that things, while seeming to behave without the involvement of a higher being, end up doing what we might expect them to do had they been guided and directed from beyond by a higher being. By a kind of fantasized or conjured up god, perhaps? A few seem also to acknowledge that the primary role of humans is to recognize the cosmos as meaningful. All in all, there are mixed feelings among scientists about things like this, and it should not be overlooked that most scientists possess little formal philosophical or theological training.

Bohm believed in "relevating" the world. To relevate something is to recognize it as relevant, to proclaim its importance, and Bohm believed it to be a primary task of everyone to relevate the world. Physicists are especially able to relevate things because nothing in the physical universe is aside from their considerations, which extend not only from soap bubbles to galaxies, as mentioned at the beginning of this book, but also down into the sub-atomic zone.

Concerned with the workings of nature, scientists more than others relevate the world from both its inner and outer behaviors. We speak, for example, of inertia as an intrinsic behavior of matter, a property whereby things tend to hold their states of motion constant, revealing an intrinsic stubbornness about them. The meanings of things, however, do not end with our definitions of them. Like everyone else, scientists are called to elevate creation into poetic dimensions of human reflection by voicing the wonder of things being what they are.

A humanizing of science, as previously cited, would require the conscious inclusion of ourselves in our discernings of nature. And because we are creatures who focus on meanings, this would demand that we speculate on meanings—on how our sciences relate to other fields of thought. This, of course, would greatly widen the scope of our studies as scientists, bringing us nearer to other fields such as philosophy, psychology, and theology. A truly broad and extensive scholarship seems to be called for, one in which a person in some way remains a broad-based student for life.

In science our goal is to know the world as much as possible "as it is." However, a growing number of scientists are recognizing that in observing the world as it is, we are attempting to know it as it is *to us!* This is not to say that scientists are falling into solipsism or gross subjectivity, but rather that some are admitting that the subjective nature, or inner forum, of observation is at least as important as the objective.

In observing physical reality we are not only observing what is "out there," but also observing ourselves connecting with what is out there. When attempting to dismiss considerations of ourselves in what we observe, we are suppressing vital data. Bohm would have argued that it is impossible to exclude ourselves when conducting observations, because subjectivity is inherent to humans.

Scientists love to think in terms of "fields," which are visualized as regions of influence. Were scientists to try humanizing science by linking it through fields of meaning with other disciplines, their efforts would be totally proper, for scientists delight in structured thinking; to them it is a highly joyous adventure.[5] Thus, professionals in the humanities must not expect scientists to cease being scientists when extending their thoughts beyond the sciences.

Above all, Bohm believed that the human brain enfolds, and perhaps generates, its own *plenum*, or fullness, of all that is, for if we are the highest known outgrowths of nature, then what happens in us should also be happening on a more primordial level in nature apart from us.

If the cosmos is to assume its rightful place in a meaningful whole within our psyche, we must understand that the world of the implicate and the world of the explicate are *interlaced*. Perhaps the scientific world will affirm this some day. Meanwhile, there are the few such as Bohm who are willing to relate their science to the innate dignity of humans who are made to know and delight in all dimensions of the cosmic mystique.

Notes

1. Christopher Bird, "The Future of the Earth against the Yoke of Contemporary Science," a paper presented at a conference entitled "Is the Earth a Living Organism?" held at the University of Massachusetts, Amherst, August 1–6, 1985. The conference was conceived by Michael J. Cohen, Director, National Audubon Society Expedition Institute, Northeast Audubon Center, Sharon, CT.

2. Michael J. Cohen, "Releasing Prejudice against Nature through Environmental Education: The Expedition Institute Model," a paper presented at a conference entitled "Is the Earth a Living Organism?" held at the University of Massachusetts, Amherst, August 1–6, 1985.

3. David Bohm, *Wholeness and the Implicate Order*, Routledge & Kegan Paul (1980), 11.

4. Ibid., p. 23.

5. Physicists have developed fascinating mathematical ways of portraying these regions in terms of their intensities, their directions, their gradients, the local spreading out or coming together (divergence or convergence) of their influences, and the extent to which they locally twist or "curl." There is a great deal of beauty in these portrayals, which those in the liberal arts would do well to study.

For Discussion

1. In what ways do you think science and art could be made to cooperate toward the goal of "re-enchanting" the world?

2. We have textbooks on "biophysics" and "geophysics." What do you think of a possible book on "artphysics"? What topics do you think such a book ought to include?

3. Classify the following as "implicate" or "explicate:" (a) a magnetic field (b) a beautiful sunset (c) an e-mail message before it is downloaded (d) a song that is being sung.

CHAPTER 28

Integrating Science and Religion

"In the beginning when God created the heavens and the earth,
the earth was a formless void and darkness covered the face of the deep,
while a wind from God swept over the face of the waters."
—Genesis 1:1–2

S et down in writing about a thousand years before the time of
Christ, the creation account in the Book of Genesis is among the
most beautiful works ever penned. It is written in a high poetic
style that was widely used in ancient Hebrew literature to express wonder in response to mystery. Readers in those days understood the scriptural accounts to be poetic. Genesis was to them not so much a literal
historical/scientific account of how God created the world as it was a
narration of that mystery. Genesis not only cited creation as wonderfully
set apart from the waste and void of nothingness; it also identified and
acclaimed the Lord God for having accomplished it.

Because the sacred scriptures are writings of transcendent faith that
were not intended to be taken as science reports, people who take them
as literal accounts confront insurmountable difficulties. We become especially aware of this when reading Genesis 1:16, which states: "God
made two great lights, the greater light to rule the day and the smaller
light to rule the night." The account states that this occurred following
"evening and morning of the third day, and afterwards there was evening
and morning on the fourth day." This chronology, of course, does not
make sense when interpreted literally.

Obviously, the word "day" as used in Genesis means something far
different from what we mean when we use it in modern science books.
We would, in fact, be hard-pressed to find a reputable scholar of Old Testament literature who interprets Genesis as an on-the-scene eyewitness
account of creation. Modern scripture scholars generally believe that the

Pentateuch, which includes Genesis, was not intended to be history in our sense of the word.

In fact, there have been numerous instances of severe misunderstandings resulting from Christians adhering to scripture passages as if they were scientific revelation. The following Old Testament passages exemplify two that were often quoted in support of heliocentricity, the erroneous view that the sun orbits an immovable earth:

> You set the earth on its foundations,
>   so that it shall never be shaken. (Psalm 104:5)

> The LORD is king, he is robed in majesty;
>   the LORD is robed, he is girded with strength.
> He has established the world; it shall never be moved.
>   (Psalm 93:1)

Clearly, these are figurative expressions suited to the mentality of the time, but not in agreement with modern astronomy and geophysics. They express wonder at the mystery of God and creation, all in relation to the universe *as understood at that time*. The Israelites of old viewed the world as having a broad, flat, unmoved earth on which rested unseen vertical columns supporting a dome-like, star-studded sky, a celestial roof of a sort referred to as the firmament. They pictured the sky above, the earth below, and the oceans below the earth (see Deuteronomy 5:8).

To some extent, Genesis presents God in the image and likeness of humans. It portrays the Lord God creating parts of the cosmos separately and sequentially, one after another in the manner that we ourselves perform acts. It describes God as walking in the garden and asking Adam and Eve "Where are you?" as if to suggest that God did not already know. It portrays God looking at creation and taking a rest as if being tired, just as we ourselves are tired after a hard day's work. Through this approach, the authors suggest the necessity of our pausing like God every seventh day—seven being a mystical number to the Hebrews—in order to reflect on meaning and refresh ourselves spiritually. And, as is repeatedly emphasized, things like that are all very good.

Some Christians today are troubled when scholars suggest that certain early scriptural accounts were derived from tales of pagan folklore, then modified into theistic accounts of praise to the one true God. They find it difficult to accept the idea of awareness of the one true God coming into human consciousness through evolving understanding. They insist that divine authorship means immediate and full revelation from God, who is imagined as having dictated scripture to writers somewhat in the manner of an executive to his secretaries, as if human understanding of the message came *after* the message was received, rather than the revelations having been gradually understood and then finally written down. They reject the notion of God utilizing the imaginations and intuitions of writers as vehicles for entering human consciousness. They seem to say that God would never use such lowly means to convey sacred messages.

Regarding the age of the earth, some Christians insist that scripture can pinpoint the creation of the earth at a few thousand years ago. They begin with the story of Cain and Abel being offspring of Adam and Eve, then follow genealogical listings throughout Genesis and other books as if they were reading literal accounts of unbroken succession down to the time of Christ. But in this endeavor they are widely at odds with the physical and life scientists.

What do scientists say about the age of the earth? As researchers of the substance of the world, they can glean information from geology and from chemistry. For example, a radioactive element with a long half-life like uranium can tell us something about the time that has elapsed since its beginnings on earth.

Geophysicists tell us that wherever uranium is found in natural deposits, the element lead is always found along with it. It is well known that uranium decays radioactively with time, ultimately becoming lead. This suggests that natural deposits may have been originally pure uranium at the time the planet was formed, and have been slowly decaying into lead ever since. Our planet, then, would seem to have within it the equivalent of a built-in clock whose zero-hour can be traced. Knowing both the radioactive half-life of uranium and the ratio of uranium to lead presently found in mines around the world, researchers can calculate the length of time that the uranium has been decaying into lead. And they suggest that this time span would correspond to the time elapsed since the formation of the earth. Utilizing this method, geophysicists attempt to calculate our planet's age, and they consistently come up with an answer of about *4.5 billion years*. In addition, radioactive materials collected from the moon suggest a similar age for the moon.

There is also abundant evidence from other sources indicating that life on earth was present very long ago. Paleontologists in British Columbia are presently studying dinosaur tracks left in cemented geological formations from at least 200 million years ago. Their knowledge of the time needed for various kinds of geological strata to be deposited enables them to make such estimates. Amazingly, if we view the estimated 4.5 billion years since the earth's formation as if it were one calendar year, then, on that scale of time, those dinosaur footprints would not have been made until about December 23rd of that year. And humans would have appeared on earth about one hour before midnight on December 31st!

These figures are vastly at odds with those of the biblical fundamentalists. In response to this, some "creationists" maintain that the world was created recently but with traces of a lengthy history. When confronted with evidence of fossil longevity, island chain formation, growth rates of mountains, and wanderings of the magnetic north pole, some insist that God created the earth a few thousand years ago *but made it appear to proud humans* as if it was very much longer ago than that. Some even suggest that God embedded fossils into the earth when the earth was new. But that view, it seems, would turn God, who throughout scripture insists that we respond to truth, into a willful deceiver or bewilderer of humans. Most current theologians, I think, would find this explanation unacceptable.

Biblical fundamentalists seem especially offended by the concept of what they call "people coming from monkeys." When quoting the sacred writings in support of immediate human creation by God, they seldom mention the passage saying that "God is able from these stones to raise up children to Abraham" (Matthew 3:9). They will admit that God raised up humans immediately from the slime of the earth, and also that God is able to raise up children from stones. But they adamantly maintain that God has definitely *not* raised up humans—rational animals— from animal substance by way of a natural process called evolution. Perhaps they have not considered that living animals, primates in particular, exemplify states of existence that are much more complex than either slime-of-the-earth or stones.

Science has uncovered many truths relating to the inner and outer workings of the world, and we must pay careful attention to those findings. We should always be careful not to polarize scientific discoveries and divine revelation. There are, indeed, distinctions to be made between the two. However, we ought not to allow the revelations of science to be seen as "merely secular." To look at the world in such a manner only nurtures discord within ourselves, causing us to view valid scientific pursuits with suspicion, regarding them as lowly or perhaps even profane while at the same time indulging in benefits brought to us by way of scientific research.

When we say today that God created the cosmos, it means much more than it did in days of old. Over the years science has given us increasingly detailed descriptions of how the earth was likely formed. However, it does not end there. People of religious faith who love science quickly translate their scientific understanding into transcendent dimen-

LEARNING HOW TO SEE

In spite of imperfections and limitations in things, one can delight in an ordinary scene such as this. Fallen pine needles form a unique pattern on segments of cracked asphalt on a city street. It is almost certain that nowhere else in creation can an exact replica of this scene be found. Appreciation of the uniqueness in things and people is part of our vision of God through the substance and situation of this world. (Photographed by the author.)

sions of God-appreciation. Scientists of religious faith do not close their eyes to the revelations of either science or religion as if to infer that they are incompatible. Always, they incorporate their grasp of the workings of the world with their faith in God, understanding themselves as fortunate to know the Creator from a multiplicity of viewpoints.

Many scientists prefer to think of earth's creation in terms of materials being drawn together, primarily under the influence of gravitation rather than having been put together in manipulative ways. There is, indeed, much value in this way of thinking. Even if the earth fell together piece by piece as if on its own, it can still be understood as having been formed by way of the Lord's creative spirit. After all, the question remains: "Who created gravitation?" People of faith come to see that what is often visualized as passive on one level of understanding is clearly active and "of the Lord" on another level of understanding.

Having considered biblical fundamentalists in this reflection, we should mention that science, too, has its own fundamentalists. Author Jack Page wrote that science fundamentalists "are bent with messianic fervor on actively rooting out every supposed superstition and ridiculing it." Such a mentality, just like biblical fundamentalism at the other extreme, only stands in the way of more holistic, inclusive views of creation.

Scientific discoveries, properly verbalized within the context of science, are never in opposition to properly formed religious discernments focusing on the wondrous. Whereas religion addresses the wonders of the spirit expressed in the context of this world, science addresses the mysteries of the world in the context of wonder. Scientists, after all, are persons who are fascinated, although perhaps very quietly, by the wondrous ways of the world. Thus, genuine religion and good science are both concerned with wondrous ways and the appeal of truth, but from different directions.[1]

It is my strong opinion that both scientists and theologians have been at fault regarding the alleged breach between science and religion. Many scientists shy away from openly discussing spirit in the context of world. While immediately experiencing their own spirit at work within their bodies, they are unwilling to discuss spirit-related notions such as revelation, discernment, morality, angels, grace, prayer, and mysticism—most of which are experienced entities which are very real, but which defy scientific measurement or computation.

Conversely, numerous Christian theologians seem wary about acknowledging the shepherdhood of matter, the ways in which matter can be credited with drawing us toward greater closeness with God. While willing to speak about God taking care of people through material substance—food, for example—they seem hesitant to articulate a materialism that is God-centered. With a few notable exceptions, such as Karl Rahner and Pierre Teilhard de Chardin, they seem unable to verbalize this well. Perhaps theologians would do well to take a few courses in the physical and life sciences so as to address on a more immediate level the fantastic loveliness of God that is to be found in the makeup of creation.

Note

1. Nobel laureate Charles Townes said in 2005: "In the long run they [science and religion] must come together." (See *New Orleans Times Picayune* section entitled "Health and Science," 11 March 2005.)

For Discussion

1. Do you find it disturbing that certain early scriptural accounts might have been derived from tales of pagan folklore and modified into theistic accounts?

2. Do you think that God would implant in a recently created earth such things as ancient dinosaur remains in order to confuse humans?

3. Do you agree that matter can be "credited" with drawing us into greater closeness with God?

CHAPTER 29

The Scientist as Mystic and Prophet

"The Christian of the future will be a mystic or he or she will not exist at all."
—Karl Rahner, S.J.

I f there is one adjective in the English language that baffles the physical scientist, it is the word *mystical*. The mystical connotes to most scientists a subjectivity that goes against the purported objectivity of all good science.

Yet it is a fact of life that no normal human can circumvent the mystical, for we cannot remain seated in objectivity to the extent of never chasing after our star. The moment we respond to what we love, we do so as persons addressing the cosmic mystique in our odysseys of meaning.

Scientists respond to the mystique of the world simply by investigating the mysterious ways in which it behaves. They utilize a kind of esoteric lingo called mathematics that is intended to reach beyond particulars and expand our understanding into the world of universals. They allow their data to hitchhike, as it were, on the wings of mathematics in hopes of describing those parts of the world that cannot be visited, or touched, or physically experienced.

Drawn by intuitions of many kinds, some scientists occasionally express their deep-seated faith in mystical principles. Some of them say, for instance, that the elegant language of mathematics best expresses the workings of nature. English physicist Paul Dirac once wrote: "It is more important to have beauty in one's equations than to have them fit experiment." Thus the mathematical scientist proceeds, always seeking symmetry, balance, and simplicity in the midst of complexities.[1]

One of the admirable qualities of scientists is their willingness to engage in self-correction, their readiness to improve on their previous descriptions of the world's workings. This quality was especially apparent in certain early American scientists such as Benjamin Franklin. Franklin is said to have amazed his European counterparts, members of the Royal

Academies, by first publishing his findings about electricity and afterwards engaging in self-correction—mentioning that certain of his previous conclusions were faulty.

Scientists generally conceal their feelings of wonder at the loveliness of what they address. Hence, they are often seen as somewhat detached and reserved, even by those who look upon their laboratories as places where impossible dreams come true. Despite the silence of scientists regarding transcendent meanings in nature, their quiet fervor in pursuing cosmic infrastructures speaks volumes. On the mysticism of scientists, Fritj Capra wrote:

> Atomic physics provided the scientists with the first glimpse of the essential nature of things. Like the mystics, physicists were now dealing with a non-sensory experience of reality and, like the mystics, they had to face the paradoxical aspects of their experience.[2]

A *New York Times* article entitled "The Isolated Scientist" began: "I always secretly hope that people won't ask me what I do. When I tell them I'm a physicist, conversation stops."[3]

In the years immediately following World War II, a golden era for scientists, physicists were viewed with awe by many. In the eyes of the public, they had accomplished what the generals and admirals had not—namely, a quick and decisive end to a heartrending global military conflict. Since those days, however, this respect for scientists has eroded. Many of today's scientists experience a kind of social isolation not unlike that of the ancient prophets.

If, like Carl Jung, we define mysticism as "entry into new dimensions," it quickly becomes evident that a certain mysticism exists in the community of scientists. With the exception of a few talented individuals like the late Carl Sagan, scientists generally hesitate to speak to the public about their visions. While harboring their love for electron spins, curl vectors, curved space, and forbidden energy levels, they understand that conversation about such things would be difficult in mixed gatherings. They stand apart from others by virtue—or, perhaps, by fault—of highly specialized knowledge that is often difficult to communicate to those lacking science training.

Prophets open themselves to the wonders of life, drawing attention to them and thereby effecting a movement toward the sacred. Scientists, too, have played the prophet by drawing attention to the wonders of creation. As persons attuned to creation's basic manifestations, they write about the marvels of the world, including the invisibles, enabling others to better understand the mysteries, to see the cosmos as very much more than it first appears to be.

The scientist evokes new awareness of matter-energy and space-time, the entities that are fundamentally as ancient as the universe. But the mysticism in which scientists have indulged has not been well communicated to others. Typically, scientists publish their discoveries in a low-key, matter-of-fact style that conceals their latent fascination with the world's behavior. Little is ever mentioned about the joys of the close camaraderie they experience while delving into the inner and outer workings of the world.

Of course, many would say that the task of effecting a transition from facts to feeling is not the scientist's, but rather the philosopher's, the theologian's, the artist's or the mystic's. Some of my own scientist friends, when asked about this, replied that they did not think it appropriate for a scientist. They seem to forget that the invitation to communicate joyous response to creation is given to everyone. Certain scientists become so privately engrossed in their labors as to greatly harm their health.[4]

In our present age, where the origin and end of the universe have become scientific questions, the languages of science and religion sometimes exhibit striking similarities. One such example is the use of expressions such as "the alpha and the omega," referring to the beginning and the end of the world. In addressing these questions, cosmologists gather together the numerous equations that express the known behaviors of the universe. These are fed into computers programmed to mathematically reach out both forward and backward in time. The question at issue is: "What is the world, by its known behaviors, telling us about itself on the grand scale of time?"

Our computers suggest that, in the case of a so-called "closed universe,"[5] the beginning may have been a sort of *zero-point fluctuation*. This is a concept strangely analogous to the scriptural notion of creation being called forth out of nothingness. In this mathematical model, nothingness is visualized by physicists as having been the right condition for matter and its opposite, anti-matter, to "burst forth" into the world out of the so-called "zero-point" during what is called "an alpha transition in the beginning."

Computers have suggested that, in the case of a closed universe, all things will ultimately proceed toward a point of singularity that scientists speak of as "the final omega point." Their ideas are surprisingly similar to the scriptural understanding of all things going back to God. They are also reminiscent of Teilhard de Chardin's intuition that everything converges spiritually toward what he called the "Omega point."[6]

Whoever listens with religious discernment will recognize spiritual overtones arising from the scientific ways of viewing nature. If, for example, in place of the word "nothingness" we substitute the expression "that which cannot be observed or measured by presently known methods," we then speak of nothingness with a kind of hope and expectation, however hazy.

With further discernment, we recognize other spiritual dimensions in the activities of scientists. Astronomers alert us to the unimaginable expanse of the universe. Electron microscopists enhance our visualizations of the innermost workings of matter. Cosmologists and paleontologists lead us back to reflections on our beginnings. And chemists give us the good example of focusing attention on materials in unbiased ways. Whereas many acclaim diamonds and deplore rust, chemists engage in evenhanded explorations of substances, as if to say that *all* atoms equally merit their attention.

Returning now to the Christian perspective of the world. . . . Inasmuch as Jesus is understood to be Truth and Love personified, it is not necessary that we literally quote him from scripture in order to communicate him to others. What is essential is that we communicate truth with love and act on it wherever we happen to be, for in so doing we are proclaiming God who is Love. What does this mean in everyday terms that are relevant to the scientific community?

It means recognizing the extraordinary in the ordinary. It means admitting that, whenever chemists teach with concern for the truth that salt is made up of sodium and chlorine atoms, they implicitly affirm the truth-value in that teaching. So it is, too, when they teach that a bond angle of 104.5 degrees exists in the hydrogen/oxygen/hydrogen configuration of a water molecule. Those who expound physical relationships such as $E=mc^2$, $a=F/m$, or $v=fl$ are likewise, perhaps unwittingly, proclaiming the presence of the Creator in their midst through the truth-expression of their formulae. A spiritually sensitive person could fittingly precede every such scientific statement with an utterance such as: "Thus saith the Lord, the God of Truth."

Mystically speaking, then, scientists who communicate their findings about matter and time can be understood as proclaiming "Thus saith the Lord about matter and time!" And similar statements can be made about all sorts of particulars. "Thus saith the Lord about energy levels and hyperfine spectra! About DNA! About nuclear magnetic resonance! About x-ray diffraction! About everything everywhere without exception!" It is equivalent to exclaiming that the Lord of Truth is present in the truth of every thing before us, behind us, above us, below us, and within us. Thus, the voicing of truth anywhere, about anything, at any time by anyone is the outcome of someone playing the prophet—usually subconsciously.

Yet it is certainly not expected of us who are scientists that we scientifically demonstrate the existence of realities beyond what we can observe and measure. It is, however, important that, when speaking as scientists, we do not deny the existence of what we cannot measure with rulers, ohmmeters, revolving mirrors, cathode-ray tubes, and the like. In asserting its own credibility science necessarily draws on recognition from realms that lie beyond itself. As expressed by Loyola University philosopher Henry J. Folse:

> ...Science must presuppose a ground of phenomena outside the phenomenal realm before it even begins its description of observed phenomena. To ask that science empirically demonstrate that such a reality exists would be to ask it to prove what...it must assume.[7]

The scriptures state that Jesus on a certain occasion uttered the words "No one goes to the Father except through me." In time it dawns on discerning Christians that all things, on levels befitting themselves, speak the same message. That is, no one in this world relates with the Creator independent of the situations and substances of this world. As humans, we go to God by way of God's creation, as wayfarers touched by grace in the twofold world of material and spirit.

Prophets, as traditionally understood, are people who listen with discernment, then communicate God's messages to others. How about scientists?

It is well known that numerous scientists have claimed to be atheists.[8] However, many are also believers. Among Nobel laureates, Albert Einstein, Paul Dirac, and Eugene Wigner believed in God, whereas Linus Pauling and Richard Feynman were professed atheists. Of those who seemed to believe in God, some appeared to do so in unusual ways. For instance, Paul Dirac once said that if there is a God, he's a great mathematician. And physicist John Wheeler, when asked if he believed in a personal God, replied that the idea of a personal God was a little too concrete for him. When asked if he thought of Christ as God, he replied, "Instead of running him down, I'd run others up." He went on to say that nature is the safest guide to questions about God, and in time the truth will be revealed to us.

Most of today's scientists do not explicitly mention the Lord as did the prophets of old, or even as the early scientists did in their writings. If scientists were to do so today, many of their listeners would certainly find this a bit "heavy" and avoid them—just as people long ago avoided the prophets, who were thought to be strange. Pending the arrival of a day when people will long to hear scientists speak of the Creator in explicit terms through the workings of the world, there is still much that scientists can do as witnesses of truth. They should try to communicate something of the wonder that resides in things, even if it be in secular language, for ours is a world where joy over the mystery of creation and delight in the Creator are ultimately one and the same.

Notes

1. An outstanding example illustrating the "beauty of alikeness" in equations is the following: Physicists long ago were able to determine the speed at which a mechanical wave travels along a stretched rope (or a string, such as a violin string). Their equation had the form:

$$d^2y/dx^2 = 1/v^2 \, d^2y/dt^2 \text{ where } v = \text{the speed of the wave.}$$

Later, it happened that Scottish physicist James Maxwell, after studying electric and magnetic fields, summed up the behavior of an electric field with the following equation:

$$d^2E/dx^2 = 1/c^2 \, d^2E/dt^2$$

Noting the similarity between these two equations, Maxwell recognized that the latter one suggested that electric field "waves" might someday be found and, if ever they were to be produced, then (by analogy with the first equation) the symbol "c" would designate their speed of travel. He solved the equation and found that the speed of those waves could be expected to be 186,000 miles/second. That was in 1862. Then, in 1886, German physicist Heinrich Hertz succeeded in producing electromagnetic (radio) waves. He measured their speed and found them to travel at the same 186,000 miles/second that had been predicted by Maxwell twenty-four years earlier. It was unfortunate that Maxwell died in 1879, thus missing by seven years seeing his predictions fulfilled.

2. Fritjof Capra, *The Tao of Physics*, Shambhala Publications, Inc., Random House (1975), 51.

3. Ronald N. Kahn, "The Isolated Scientist," *New York Times Magazine* (January 5, 1986), 38.

4. The following illustrates a personal experience in this regard. Early in 1949, I began employment as a vacuum tube engineer with Farnsworth Radio and Television Corporation. Fresh out of Notre Dame's Graduate School of Physics, I drove to Ft. Wayne where Philo Farnsworth, the great inventor of television, conducted my job interview. Sixteen months later I arranged a job interview in Ft. Wayne for my friend, Notre Dame machinist Robert Lyvers, who immediately became Farnsworth's personal machinist. Lyvers later related how Farnsworth, caught up in his laboratory work, la-

bored extremely long hours to the detriment of his health. He died in 1971 at the age of sixty-four.

5. A closed universe is one in which the speed of expansion of its components is low enough that gravitation will eventually bring them back together in a future cataclysmic collision.

6. The final chapter of this book addresses the "Omega point."

7. Henry A. Folse, *The Philosophy of Niels Bohr*, North-Holland (1985), 243.

8. See, for example, Denis Brian, *Genius Talk, Conversations with Nobel Scientists and Other Luminaries*, Plenum Press (1995), 28, 49, 69, 133.

For Discussion

1. Does the idea of matter and anti-matter "bursting into being from a state of nothingness" disturb you?

2. Is it possible for a scientist to work as a scientist without implicitly acclaiming in some way the spiritual realm, such as that of "the wondrous" or "the worthwhile"?

CHAPTER 30

Science and the Miraculous

*"There is a profound truth to the statement
that all questions are ultimately theological."*
—Sean P. Kealy

A ges ago religious people regarded rainbows as the product of preternatural activity on the part of God, who was envisioned by many as an artist painting on a canvas. The fact that rainbows are strikingly beautiful and are specifically mentioned in the Old Testament surely contributed to these sentiments. The Genesis account reads:

> "This is the sign of the covenant that I make between me and you and every living creature that is with you, for all future generations: I have set my bow in the clouds, and it shall be a sign of the covenant between me and the earth. When I bring clouds over the earth and the bow is seen in the clouds, I will remember my covenant that is between me and you and every living creature..." (Genesis 9:12–15).

It seems apparent that the Israelites of old came to visualize rainbows in ways that were different from the ways in which they had been previously understood. That is to say, *a shift in awareness* seemed to have occurred at one point whereby rainbows were subsequently viewed as direct expressions of the Lord's special concern for them.

Questions arise: Now that scientists understand rainbows in terms of refraction, dispersion, and reflection of sunlight, and now that the mystery of their colorings has been solved, are we to think of the scriptural account on rainbows as presently less valid than it was in Old Testament days? Does a covenant no longer exist between God and people, a covenant of which rainbows were/are the sign?

Some people might answer this in the affirmative, stating that modern science has dispelled old myths and superstitions. But an alternate view would be that not only rainbows but *all of creation* is a sign of God's special concern for humans, for everything in creation is of special significance to those who know how to see.

In our present world, where theologians write of God in terms of "holy mystery,"[1] it is not surprising that people would think of nature's mysteries in terms of God. It is psychologically satisfying to do so, for nature consistently leads us to the edges of understanding, repeatedly bringing us to face the beyonds of human observations. And from there, some persons choose to go forward in their odysseys of meaning. Others back away from delving into meanings except on a superficial level.

One area in which science and religion are sometimes believed to be in opposition to each other has to do with miracles. In a world where scientists are concerned with nature's consistent behavior as apparently governed by the so-called "physical laws," theologians have often referred to miracles in terms of God's *suspension* of physical laws at certain moments. Such statements, of course, invite raised eyebrows on the part of physical scientists.

The dictionary defines a miracle as "an event in the physical world deviating from known laws of nature, or transcending our knowledge of these laws." A less formal and more subjective definition might be simply "a wonderful thing or event." I will employ both meanings in the following account, hoping that the distinction will always be apparent.

Numerous scientists would object to our first definition and deny the possibility of miracles as events with preternatural causes. But if we speak of the miraculous subjectively as a wondrous event, employing the second definition, most scientists would probably have no problem with it, for the training of scientists opens their minds to the recognition of nature's wonders, including many marvels that non-scientists fail to recognize. Scientists are fascinated by such wonders.

When considering the miraculous, we should bear in mind that the concept itself has changed in certain ways across the ages. For example, in the Old Testament account of the Tower of Babel (Genesis 11), the proud Mesopotamians were said to have built a tower so high that it "reached into the heavens." In a sort of retaliation, the Lord was said to have confused them, so that even the words of their learned men could not be understood. The scriptural writer implied that God's response on that occasion was in the nature of a miraculous intervention.

Based on today's understanding, one scientific explanation of the Tower of Babel might hinge on the idea of multiple echoing. It seems quite likely that baked tiles, made of materials native to Egypt, were used for the inner walls of the tower, violating what we now know as principles of good acoustical design. Lingering echoes might have made it difficult for speakers to be understood. The effect, known as reverberation, is well understood today.

The Old Testament story of the fall of Jericho also comes to mind. A miraculous event is described in which the walls of Jericho, a city filled with "God's enemies," came tumbling down when the Israelites shouted and blew trumpets. To the populace in those days, such an event seemed preposterous. On the basis of today's knowledge, however, one might explain the Jericho miracle as the result of seismic activity at a superbly opportune moment, for it is known today that Jericho lies on a geological fault line, and that throughout history its walls have fallen numerous times from earthquakes.

There are, of course, Old Testament miracles that are not so easily explained. One involves the account in the Book of Joshua of the Lord having made the sun stand still for a time. It was said that the Lord did this so that the chosen people might win their battle against the Amorites, who were regarded as enemies of God. During the battle, darkness was approaching, so the story goes, and additional daylight was needed. The account reads:

> On the day when the LORD gave the Amorites over to the Israelites, Joshua spoke to the LORD; and he said in the sight of Israel,
>   "Sun, stand still at Gibeon,
>   and Moon, in the valley of Aijalon."
> And the sun stood still, and the moon stopped,
>   until the nation took vengeance on their enemies.
>   (Joshua 10:12–13)

Numerous questions arise: Might it have been that the region was experiencing overcast weather for some time, causing night to fall earlier than usual for several days? Might it then have happened that, at precisely the favorable moment for the Jewish soldiers, the clouds parted and the sun shone through? This would have *in effect* extended daylight, so that the sun would have appeared to stand still, in a manner of speaking. According to a recent observer, archaeologist J. B. Pritchard, the shape of the valley and the presence of dust in the west often prolong the effect of sunset in that region, thus seeming to extend the hours of daylight. But the scriptural account also says that the sun halted in the middle of the sky:

> The sun stopped in mid-heaven, and did not hurry to set for about a whole day. There has been no day like it before or since, when the LORD heeded a human voice. (Joshua 10:13–14)

This passage is far less easy to explain as a natural occurrence than the preceding ones, and some poetic enhancement should not be ruled out. Certain modern scholars perceive the entire narrative as a fictionalized version of an actual battle, a historical account with some poetic license taken. Nevertheless, the entire account was destined to enter a conflict thousands of years later when Galileo attempted to convince church authorities that the earth orbits the sun.

Among the greatest of Old Testament miracles was the parting of the Red Sea. The account reads:

> Then Moses stretched out his hand over the sea. The LORD drove the sea back by a strong east wind all night, and turned the sea into dry land; and the waters were divided. The Israelites went into the sea on dry ground, the waters forming a wall for them on their right and on their left. (Exodus 14:21–22)

Was this phenomenon a miracle in the sense of nature's consistency of behavior being suspended? Did it appear in the manner depicted in Cecil B. deMille's film *The Ten Commandments*? Or was it an ordinary event, although one of wondrous proportion in the consciousness of the Israelites? Did those ancient people simply view it in the highest possible way because of its special meaning to them? After all, we do find poetic accounts in the Old Testament of such things as hills dancing and trees clapping their hands.

Numerous scripture scholars today believe that the Hebrews did not actually pass through a sea as ordinarily understood. Instead, the speculation is that the Hebrews passed through a shallow and marshy region with reeds—a "sea of reeds." There are known inlets, even today, along the northern end of the Gulf of Suez. We might reasonably assume that strong winds of dry air, combined with tidal forces, pushed waters out of the marshes for longer than usual periods. Scholars also point out that people may have been better able to move through marshy regions on foot rather than with horses and chariots.

Was the Exodus, then, a miracle in the sense of God actively coming to the aid of the people with an event that would not have otherwise been possible? Or was the parting of the waters an ordinary occurrence that just happened at a superbly fortunate moment? Undoubtedly, the Israelites viewed it as miraculous while the Egyptians saw it as a natural calamity that just happened to favor the Israelites.

In response to this consideration, Robert Gnuse writes:

> What would we have seen had we been there? Some of us with a penchant for debunking might say, "There is nothing miraculous

about this. Moses was aware of the regularity of this climatic condition, bizarre though it might appear, and led the pursuing Egyptians into a trap." Others of us, more aware that the perception of divine activity is always an act of faith, might say, "How wonderful that God can act thus through the regular course of natural and historical events."[2]

Tremendous coincidences do sometimes occur. Writing subjectively for the moment, I refer to the greatest coincidence I ever experienced. In 1958, as I was driving along an isolated gravel road, a small stone kicked up by a passing car produced a circular nick on my otherwise flawless windshield. The nick was about the size of a dime. Two weeks later, while driving along a paved highway with an engineer friend, I began telling him about how I had acquired that minor damage on my windshield. In my explanation, I reached out to point to the little nick that was situated just below the rearview mirror.

Exactly at that moment, a large gravel-hauling truck was approaching. A stray object—perhaps a heavy rock—flew out from a flapping tarpaulin atop the truck. Its flight path was such that it struck dead center on the tiny nick. So there, before our astonished eyes, the nick on the windshield erupted to the size of a silver dollar *at the exact instant* that I touched it from inside the car! There was, of course, much laughter over this extremely rare occurrence.

Amazing as it was, the event was certainly explainable in terms of people and things interacting in ordinary ways. Hearing about it afterwards, certain friends of ours seemed unimpressed. However, my passenger and I viewed it as especially wondrous, since our science backgrounds enabled us to recognize the staggering odds against such an occurrence. The moving automobile was in exactly the right place, at exactly the right moment as the randomly ejected flying object struck that exact spot on the windshield at the very instant my finger came in contact with it. [Note: This account is written here literally and with absolutely no enhancement].

Modern theologians typically maintain that apparently miraculous events should, wherever possible, be explained first on the basis of natural causes—through God's ordinary way of accomplishing wonders. They believe that in most cases we are dealing with ordinary occurrences that are afterwards seen in a transcendent light. Some point out that spiritually sensitive people are more likely than others to regard *all* of nature as a kind of miracle or wondrous phenomenon.

The miraculous in things, places, and events becomes apparent to those who, as Gregory Baum writes, "already acknowledge the marvelous as a dimension of human history."[3] Seemingly, those who are receptive and willing to have their eyes opened are the ones who see. This principle seems to lie at the root of religious faith in general.

Many wondrous events occur privately within people. One such happening was related by theologian Joseph Grassi as follows:

> Years ago, I was alone, seated in a room with closed doors. Suddenly, I perceived the room to be shaking. Thinking it was an earthquake, I got down on the floor and felt the wall which was also vibrating. I then made my way to the door and entered the next room where there were several people. I asked them what they thought about the earthquake. They told me that they had not felt the slightest tremor, but they then announced that they had just heard an interruption on television describing how President Kennedy had been shot moments before. In reflecting on this experience I can only hypothesize that somehow I was in tune with this event that shattered millions of people.... Somehow I had been in a state of consciousness that transcended the barriers of time and distance and tuned in to happenings I ordinarily could not perceive.[4]

Grassi's experience suggests the possibility that the earth tremor at the death of Jesus might have been perceived by some and not by others.

Are miracles, then, to be understood as phenomena that are purely subjective? Are they never to be thought of as exceptions to the physical laws? The answer hinges, at least in part, on the extent to which our formulation of physical laws is truly all-inclusive. It is certainly possible for scientists to prematurely formulate statements of nature's consistent behavior, to do so before sufficient evidence has been accumulated. We could argue that an infinite duration of time would be needed for formulating laws that are totally all-inclusive. So, if we recognize our physical laws as tentative even to a slight degree, we thereby admit the possibility of exception to these laws as formulated. Such an admission would become a factor enabling scientists, as scientists, to reasonably allow for miracles, seeing them as events occurring beyond the *ordinarily observed* workings of a not-yet-fully-investigated natural world.

As data-conscious persons, scientists cannot logically deny the possibility of exceptions to the formulated laws of nature. For us to hold such views would imply that we possess the data that tells us so. Obviously, we do not possess such data. Nature as we know it thus far seems never to yield that kind of information about itself. Nature's behavior tells us very much about how it behaves, but far less about how it cannot behave.

Fundamental questions remain. Might it be that a characteristic of rarity or novelty resides in every thing and every event that has ever occurred in the whole of cosmic history? In a cosmos of inestimable dimension in which the presence of matter is the exception rather than the rule, might it be that the whole world possesses a quality that we might grow to recognize as miraculous? Might it be that we are in some way indulging in the miraculous in every place and at every moment, even while speaking of our worldly experiences as nothing beyond the ordinary? Is that which we call "ordinary," when considered from a truly large perspective, really "extraordinary?" There are many highly spiritual persons, I think, who would answer these questions in the affirmative.

As for bona fide miracles involving radical departure from what we call "natural laws," do such things take place? I, for one, am convinced that they do. Some of them have been so well documented that an impartial observer would not reasonably deny their occurrence. The many documented cases of medical cures at Lourdes in France come to mind. Instant cures, such as those of people with advanced stages of cancer, have been documented by x-rays, sometimes by doctors who went there to scoff.

The New Testament abounds with examples of conversions of the human heart wrought through the workings of Jesus. Repeatedly, Jesus demanded faith as a prerequisite.

We easily demand miracles as a condition for our wholehearted belief in God while failing to see that we ourselves, rescued from nothingness and given life, exemplify the miraculous. Jesus seemed unwilling to perform miracles for persons who demanded miracles:

> The Pharisees came and began to argue with him, asking him for a sign from heaven, to test him. And he sighed deeply in his spirit and said, "Why does this generation ask for a sign? Truly I tell you, no sign will be given to this generation." (Mark 8:11–12)

It is of interest to note that mystics through the centuries have often referred to Jesus himself as the Sign Unrecognized.

Miracles of inner conversion do take place. A person suddenly turns from hatred to love, or experiences inner peace when a painful memory is healed. Often these inner miracles are accompanied by easily observable signs. The eyes of a happy person are very different from those of a vengeful person. So, too, is the energy level. These physically observable differences lead us to ask with genuine interest, "What happens within a person, sometimes suddenly, that causes these externally observable signs to occur?" Certainly the answer lies beyond the usual analyses of the physical sciences.

A truly great miracle in Christian consciousness is that in which we grow to recognize ourselves and others in terms of Christ being personally at work in the world through his followers. Christians, in time, come to see themselves as literally—not just figuratively or symbolically—

advancing the messiahship of Christ by way of their daily engagements. When serving others, they understand it to be the Lord Jesus who is serving others through them. When they suffer, they fathom that it is the Lord Jesus who is suffering in this world through them. And when doing what they do with the good of others in mind, they understand themselves as enabling Christ to continue his messianic role in today's world through their hands, their mouths, their legs, their minds.

This conviction is easier spoken about occasionally than lived on an hour-by-hour basis. Living it demands moment-to-moment awareness and immediate consciousness of personal identification with Jesus the Lord as we load the dishwasher, adjust the thermostat, answer the telephone, cultivate crops, bury a dead horse, or bear with the eccentricities of others.

In Christian eyes the greatest miracle ever to occur was the Incarnation wherein God personally identified with creation so as to make everything understandable in terms of divinity, and divinity understandable in terms of everything everywhere. The ineffable miracle is that of having our eyes opened so as to see that ordinary creation is alive with the grandeur of love-expression, forgiveness, and the promise of everlasting life. Whoever believes in things like these is already living the miraculous.

Notes

1. Karl Rahner, S.J., writes: "We want to call the term and source of our transcendence 'the holy mystery', although this term must be understood, deepened, and then gradually shown to be identical with the word 'God'..." (*Foundations of Christian Faith,* trans. William V. Dych, S.J., Seabury Press, [1978], 60–61).

2. Robert K. Gnuse, with Charles Winters, *The Jewish Roots of Christian Faith*, New Orleans, Loyola University Institute of Ministry (1983), II-8.

3. Gregory Baum, *Man Becoming: God in Secular Language*, Herder and Herder (1970).

4. Joseph Grassi, An unpublished account in the hand of the writer. Grassi is author of *Changing the World Within*, Paulist Press (1986), and other publications. He is a professor of religious studies at Santa Clara University, San Jose, CA.

For Discussion

1. Do you have any personal "rainbows" that you regard as special signs given to you by the Lord?

2. Name one or more things in your life that you consider to have been miraculous.

CHAPTER 31

On the Mystery of Time

"What then is time? If no one asks me, I know.
If I want to explain it to a questioner, I do not know."
—St. Augustine

"Time is the great mystery of nature
which keeps everything from happening at once."
—C. J. Overbeck

I clearly remember an instance more than half a century ago when I had to memorize a definition of time dating back to the ancient Greek philosophers. "Time," it said, "is the measure of motion in the sense of before and after in a process." It is a curious definition that has haunted me to this day.

Bernard Lonergan refers to time as "the ordered totality of concrete durations."[1] Both his definition and that of the ancient Greeks are enshrouded with mystery. Time itself, and the words that we commonly use to describe it—words like before, after, moment, instant, duration—are all part of a very broad, very deep mystery.

Most people believe strongly in the concept of instantaneous time, a kind of all-pervading present moment encompassing the whole of the universe. According to this belief, we on earth may say to ourselves, "I wonder how Polaris, the North Star, looks in the clear sky tonight!" But because Polaris itself is seventy-five light-years away (meaning that it takes seventy-five years for its light to reach our planet), we always see that star as it was seventy-five years ago. Thus, to put this another way, we can go outside on the night of our seventy-fifth birthday, face north, and exclaim, "Now I see you, Polaris, just as you were at the time I was born!"

Peering outward from ourselves, we perceive nature not as it is in the absolute, all-pervading instant we call "now." Rather, we see it in a relative way as extended across the past. This means that whenever we study the stars, it is not only the fascinating distant glitter of jewel-like objects that is at issue. There is, additionally, the time-spread dimension of nature at which we are gazing from our locale. Across the nighttime sky, we are looking into different depths of antiquity in a single glance. The farther away the observed stars are, the deeper into antiquity we are gazing. It is as if the pages of history on a cosmic scale are open to us in different directions all at once.

For example, let us imagine a lady standing on a planet in another solar system, a little more than five hundred light-years away from the earth. She is looking at the earth through a super-powerful telescope. As she zooms in, her lens focuses on the three ships of Christopher Columbus approaching the wilderness that will come to be known as America. That event from our distant past would be seen by her as taking place in her present moment because her observations of America's discovery are shifted by more than five hundred years from ours.

It seems that very few people are aware of the privilege that is theirs when they gaze at the stars. If only we had telescopes powerful enough to obtain detailed views of planets in other star systems, what views of ancient happenings might be ours!

Considerations such as these also hold true for things near at hand. For example, considering the time it takes for light to travel from objects to eyes, we could validly argue that we never see even our own hands or feet as they are right now, but rather as they were some trillionths of a second ago. There is a further delay because of the interval of time between the stimulus received and the awareness that follows, for there are time delays within our nervous system that hinder instantaneous awareness.[2] Such limitations of the human sensory process are of concern to both physical and life scientists.

A scientist cannot speak about motion without implying time, or about time without implying motion, such as the motion of our planet on which our units of time have been based for centuries.[3] A system of time based on the motion of the earth seems to imply that time exists in nature. Then again, we can also hypothesize a kind of eternal time that flows on endlessly, independent of any thing or person, with things and people entering it, later to exit it or assume new forms. Indeed, it is not an easy task to decide whether nature exists in time or time exists in nature. Perhaps they may be best understood as two facets of a single reality.

At any rate, in ages past, time was largely viewed as cyclic, and even in our present era certain Eastern religions view life in terms of repeated earthly reincarnations. In Western cultures, however, we see little reason for believing that time is cyclic, or closed like a loop.

Does it make any difference whether we view time as cyclic? Indeed, it does. The Stoics of old believed in cyclical time. They visualized events in the future to be repetitions of events in the past. Consequently, they believed in determinism. In their view, there was little that one could do to bring about real change in the world,[4] and their lifestyles often reflected that belief.

What might it be like to view time as cyclic? From my experience as a college professor, the academic world is a place where time is experienced as somewhat cyclic. After all, to a faculty member teaching the same courses year after year, academia is essentially a world where last year's performance, hopefully improved, becomes this year's performance, with new students beginning at the same level as last year's students. Perpetually, the students remain young, and so it is easy for a professor to become caught up in these annual cycles among the young. My own experience was that, as decades went by, I felt myself to some degree losing touch with the overall linear passing of time until I visited old friends away from campus or surprised myself in a mirror.

The ancient Israelites generally believed in the linear or straightforward flow of time. However, we do find occasional instances of an implied belief in cyclic time, at least in the case of a few writers like the author of Ecclesiastes, who wrote:

> What has been is what will be, and what has been done is what will be done; there is nothing new under the sun. (Ecclesiastes 1:9)

Following the death of Jesus, Christian belief about the crucifixion and the second coming has mostly been expressed in terms of linear, non-repetitive time. The common understanding is that the crucifixion occurs but once, and that Christ will come again, but in a new, different, glorious, and altogether lasting way.

Scripture is largely concerned with events across the course of history that signify God's loving concern for humans. It recognizes God as active on our behalf at every moment. Science, meanwhile, is not directly engrossed with this view of time. Instead, it regards time as pertaining to durations and outcomes in the interplay of things. Scientists are mainly interested in time as expressive of such things as motion, sequence, slowness or rapidity of progression, simultaneity, the duration it takes for light to travel across the diameter of an atom, and the time needed for radio signals to reach a space probe approaching Jupiter.

Isaac Newton believed in the existence of absolute time—a kind of ruthless forward movement of the present "now" that pervades the whole of the cosmos, such that an hour is an hour simultaneously everywhere. According to Newton, all motions may be accelerated or retarded, but the flow of absolute time moves at an even pace and is not subject to change.

Modern physical scientists, on the other hand, visualize time in two ways: in ordinary ways as seen in everyday occurrences, and in relativistic ways involving entities moving at super-high speeds approaching that of light. They tend to shift from one view to the other as needed to best express what they are trying to understand and describe.

In the relativistic view formulated by Einstein in 1905, space and time are visualized as a single entity called space-time, and time's rate of flow depends on the states of motion of different observers relative to what is being observed. Einstein believed that matter tends to distort space-time, so that gravitation is understood not so much as a force acting between objects, but rather as an outcome of space-time being distorted or "curved."

Certain physicists today are proposing that time, like matter and energy, is made up of particles. Theirs is not an unreasonable theory, inasmuch as matter and energy, once perceived as continuous, have now been shown to be granular in nature, existing as discrete atoms and photons. They suggest that the duration of the proposed "chronons" of time is 10^{-43} second or shorter,[5] so extremely brief as to be yet undetectable. They speculate that chronons serve in some strange way to "keep things going."[6]

We can imagine the excitement within the scientific community if chronons are ever proven to exist. Philosophers, too, will have a field day reinterpreting human experience from the standpoint of personal involvements in a realm of chronons.

As for the notion of a universal all-pervading present moment, to which some writers refer as "the subjective now," I do believe in this concept myself. I tend to identify it with the nature of God, who is immediately present everywhere at once. Yet I also believe in Einstein's relativity. Is there a contradiction between these two—i.e., between the understanding of God as an absolute who is immediately present everywhere at every moment and the understanding of time as a relative entity? I would say no, for the two perspectives view the world in apparently truthful ways but from different vantage points.

When speaking of time's relativity, physicists love to cite Einstein's fascinating example of the so-called "twin paradox."[7] In accord with this theory, we might imagine a man and his wife, both age twenty, standing at a rocket-launching site and bidding each other farewell. The husband takes off in the rocket and travels far away at speeds approaching that of light (a physical impossibility by today's technology). Upon returning he finds himself, because of the particular high speeds at which he traveled, to be only twenty-one years old, whereas his wife is now forty-three! Both have aged, but not at the same rate. Not only their bodies, but also their clocks disagree by twenty-two years.

To what extent can we recognize this example as realistic? Strange as it may seem, this phenomenon, known as *time dilation*, has been tentatively verified by actual experiment[8]—although not with people or other living creatures. Rather, it has been verified, at least initially, with the use of super-accurate cesium clocks. These clocks, after being precisely synchronized, later disagreed by a very minuscule part of a second after one clock was made to travel very fast compared to the other. Strange indeed, but true!

Reflective persons tend to regard time in terms of various events and personal associations, such as good times, hard times, old times, happy days, defining moments, moments of truth and such. As noted in the previous chapter, there is no denying that we are highly impressed by the phenomenon of coincidence. Undoubtedly, one of the ways in which the Spirit speaks to us is by way of coincidence recognized. Michael Shallis writes:

> Coincidences enable a new perspective on time to be perceived, and a new perspective on the world revealed.[9]

Psychologically, our ongoing awareness of past, present, and future viewed together as a single interrelated entity is of crucial importance. A magazine article entitled "Timeless Minds," addressing the subject of post-hypnotic suggestion, described the following experiment conducted with college student volunteers.[10]

Under hypnosis, three groups of volunteers were told that, when they awoke, the past, the present, or future would be gone out of their lives. Upon awakening, those who were told that the past would be lost became infantile, egocentric, and inhibited and suffered loss of memory. Those who were told that there would be no future experienced a loss of identity and were free of both anxiety and motivation. The ones most devastated were those who were told that there is no present. They became depressed and almost schizophrenic.

Another group, made to feel as if the future were expanded, experienced fulfillment, serenity, and no fear of death. Sentiments such as theirs are sometimes said to be experienced by people who have come very close to death.[11]

The article went on to suggest that memory of past experiences, awareness of present experience, and anticipation of future experiences are necessary for our psychic health. It implied that fulfillment, expectation, and a sense of direction in life are vital time-related psychological forces. To put it another way, our odysseys of meaning are closely related to our understanding of the "befores" and "afters" that we associate with our experiences of the present.

Is it fundamentally correct, then, to think of the past and the future as entities in their own right—as apart from matter-related things and events? Or would it be more proper to regard the past and the future as mental constructs, as notions that do not stand on their own outside of our minds that normally recall the past and plan for the future?

There are well-known opinions that the latter is true, that humans enjoy a sense of time's passage that is distinguishable from the passage of time in the world outside of themselves. Science philosopher Alfred North Whitehead wrote:

> So far as sense awareness is concerned, [the human] mind is not in time or in space in the same sense in which the events of nature are in time, but ... it is derivatively in time and in space by reason of the peculiar alliance of its passage with the passage of nature. Thus, mind is in time and in space in a sense peculiar to itself.[12]

Whitehead went on to explain that nature apart from humans is responsive to the instantaneous present, for with nature the past is over and the future is not yet. But this is not the case with humans. The human mind is a uniter of time; it reaches out and draws time inward to itself. In the human mind, Whitehead says, past and future "meet and mingle meaningfully" in the present.

Consciousness of continuity is a vital resource within ourselves. Continually, we draw upon memories from the past when planning for the future. Often we envision our future in terms of improvements over our past performances, viewing the past, present, and future as interwoven in the singular experience of life moving forward.

Seemingly, the past and the future are mental constructs seated in our consciousness and imagination, and they function to embellish our present experience by giving it higher meaning. The same might be said of other time-related phenomena, such as precognition, premonition, clairvoyance, time displacement and dislocation, time contractions and expansions, hauntings, time-slips and other time-related phenomena.[13]

Our understanding of time certainly constitutes an extremely powerful psychological factor, especially in the minds of imaginative persons who readily consider the ultimate consequences of time's passing. Let us recall Carl Jung's statement that the psychological factor of greatest power in our thinking is the god.[14] It becomes easy, then, to see why time is so often mentioned in relation to God. This is not to suggest that time itself is divine. But in the realm of human transcendence it is a fact that consciousness of time consistently appears in liturgical prayer. Expressions such as "world without end," "now and forever," "in the beginning, is now and ever shall be" all attest to this. Simply put, we find it appropriate to express life's most profound hopes in terms of time's unabated progression leading us toward a grand fulfillment. Karl Rahner wrote:

> Time becomes madness if it cannot reach fulfillment. To be able to go on forever [in this world] would be the hell of empty meaninglessness. No moment would have any importance because one could postpone and put everything off until an empty later which will always be there.[15]

Rahner explained that we cannot logically picture eternity in the same framework we use for time, for eternity is not an endless progression of time. Eternity does not come *after* our experience of time in our biological life, as is often imagined. Rather, eternity is *already with us and subsumes time*. It includes time as a particular under a universal. In our present mode of existence, eternity is actualizing itself toward its fulfillment. It remains in us after time slips away. Eternity becomes the crowning mode and accomplished victory of our existence where, following our exercise of freedom/choice in time, our *becoming* ceases and our *being* endures.

Time is a profoundly great mystery that defies definition. We who use it well seem to sense ourselves as managing it. And, when growing old, we sense ourselves to be "losing our grip" on time. No one can fully say what time is, or how it began, or in what manner it will end if, indeed, it ever ends in the nebulous way in which we visualize things as ending.

Time is like an ever-beating drum that won't slow down. It is said to "lower the hills" and "raise up the valleys." Its messages of meaning are heard, but in different ways by different persons. Once in a while time appears to accelerate: a baby seems to become a teen faster than expected, or a leap year seems to arrive every other year, especially as we get older. Time beckons us. It overtakes us. It takes its toll. It makes things possible. It hinders us from doing many of the things we would wish to do. It eventually renders us weak and brings our bodies to inglorious endings.

We lose time. We find time. We make up for lost time. We think of time as being elastic, as being stretchable so that it can include more. Our awareness of its constraints lures us to do certain things "while there is still time."

Time is said to betray us by stealing from us our youth, or by taking away our old friends. It is the mystery about which Edward Fischer writes:

We are still ... realizing more each year how completely everyone
is alone, as isolated as a star in space. Many of those who helped
us to forget our aloneness have stepped outside of time and let
the work of the world go on. As the ritual of the requiem grows
repetitive, we who are left are reduced. Seeing the many crossed-
off names on our Christmas mailing list, and aware that we are
living on the outer edge of Nature's permission, we find each day
a dividend.[16]

Time is the "familiar stranger"[17] that heals, sooner or later allowing
physical pain and unpleasant memories to fade. It offers us repeated new
beginnings, not unlike the forgiveness offered to us by the Lord. Ultimately, it is an enabler that allows God's final judgment to overtake us.
Please, Lord, with forgiveness!

Notes

1. Bernard Lonergan, S.J., *Insight: A Study of Human Understanding*,
Harper and Row (1957; paperback version published in 1978), 143.

2. Alfred N. Whitehead, *The Concept of Nature*, University of Michigan
Press (1957). Sense awareness and time are discussed in detail in chapter 3.

3. In 1967, the duration of a second of time was redefined in terms of
the radiation emitted by excited atoms of cesium-133. But the concept of
second dates back to ancient considerations of the apparently unvarying motion of the earth.

4. Richard Morris, *Time's Arrow: Scientific Attitudes Toward Time*, Simon
and Schuster (1984), 21.

5. This turns out to be less than a billionth of a billionth of a billionth
of a billionth of a second.

6. Morris, *Time's Arrow*, 21.

7. An easy-to-read account of the twin paradox appears in several edi-
tions of Paul Hewitt's *Conceptual Physics*, Little, Brown (1985).

8. The verification occurred in 1976 at which time a cesium clock was
flown around the earth, then compared with another left behind. Verifica-
tions have also been conducted with shifts in wavelength of emitted radiation from iron-57 undergoing centrifuging.

9. Michael Shallis, *On Time: An Investigation into Scientific Knowledge
and Human Experience*, Schocken Books (1983).

10. Robert A. Freitas, Jr., *Omni Magazine* (February 1984), 38.

11. For examples of this, see Raymond A. Moody, *Life After Life*, Ban-
tam Books (1976).

12. Alfred N. Whitehead, *The Concept of Nature*, University of Michigan Press (1957; originally published in 1920), 69–70.

13. These topics are treated in Michael Shallis's *On Time*.

14. Carl G. Jung, M.D., *Psychology and Religion*, Yale University Press (1938), 98. "That psychological factor which is the greatest power in your system is the god, since it is always the overwhelming psychic factor which is called god. As soon as a god ceases to be an overwhelming factor, he becomes a mere name. His essence is dead and his power is gone. Why have the antique gods lost their prestige and their effects upon human souls? It was because the Olympic gods had served their time and a new mystery began: God became man."

15. Karl Rahner, S.J., *Foundations of Christian Faith*, trans. William V. Dych, S.J., Seabury Press (1978), 271.

16. (University of) *Notre Dame Magazine* (Spring 1987), 50.

17. J. T. Fraser, *Time, The Familiar Stranger*, University of Massachusetts Press (1987).

For Discussion

1. In your personal experience, what is the most gratifying thing, and what is the most perplexing thing about the mystery of time?

2. Relative to the above account about the hypnotized students, were the reactions of the three groups what you would have expected?

3. How do you imagine yourself, a creature in time, already experiencing something of eternity?

CHAPTER 32

Is Christ the Meaning of Time?

"Very truly, I tell you, before Abraham was, I am"
—John 8:58

In a typical classroom discussion on the subject of God and time, some instructors begin by sketching a circle on the blackboard. The circumference of this circle is said to represent the flow of time—past, present, future—along which humans are progressing. Students are asked to visualize God as dwelling at the center of the circle, such that God can look

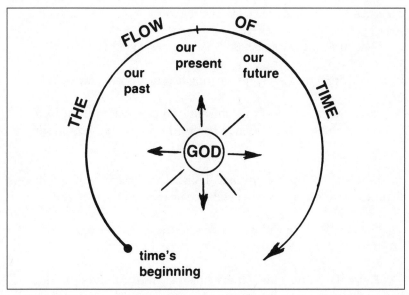

A diagram often used when explaining that God lives outside of time whereas we live within the passage of time.

outward and see our past, present and future equally and all at once. In this manner, the instructor explains that God sees all of time as if in a single and immediate glance.

How can one go so far as to suggest that Christ gives time its meaning when time itself can be traced back for eons before Jesus was born? Let us begin by considering what the gospel writers said about Jesus in relation to time.

The writer of John's gospel linked Jesus with time in the highest ways imaginable:

> In the beginning was the Word, and the Word was with God, and the Word was God. He was in the beginning with God. All things came into being through him, and without him not one thing came into being. What has come into being in him was life, and the life was the light of all people. The light shines in the darkness, and the darkness did not overcome it.... The true light, which enlightens everyone, was coming into the world. He was in the world, and the world came into being through him; yet the world did not know him. He came to what was his own, and his own people did not accept him. (John 1:1–5, 9–11).

According to the gospels, what did Jesus say about himself in relation to time? First of all, when speaking of the past, Jesus steadfastly insisted (to the point where people almost stoned him for it) that he had coexisted with his Father even before the world was made:

> "Very truly, I tell you, before Abraham was, I am." (John 8:58)

It is also written that Jesus, anticipating his crucifixion, prayed:

> "So now, Father, glorify me in your own presence with the glory that I had in your presence before the world existed." (John 17:5)

In relation to the present time, Jesus solemnly promised his followers that he is with them as their ongoing companion in the here and now:

> "Remember, I am with you always, to the end of the age." (Matthew 28:20)

As for the future, Jesus' ultimate message was one of resurrection and life everlasting—an endlessly expanded state in which he, his Father, the Holy Spirit, the angels and the saints will dwell together in glory:

"This is indeed the will of my Father, that all who see the Son and believe in him may have eternal life; and I will raise them up on the last day." (John 6:40)

In recent years, a new awareness of time's meaning is developing with regard to the Christ-event. This is not so much a notion of God breaking into history and placing Jesus in the world at a particular moment. Nor is it about God's presence entering creation where that presence was previously absent. Least of all is it an awareness of the physical nature of time shifting in some mysterious way.

Rather, the new consciousness views time as inherently sacred because it has contained the Eternal One in an unprecedented way. It recognizes the cosmos as having been imbued with God's presence from the beginning, from the very onset of creation. It sees evolution and human activity as examples of the world's makeup responding to God's power, although not fully. It sees Christ as the one who responded fully to the presence of God in our world, awakening us to that reality so that we might follow his lead. And because of Christ's presence having sanctified time, Christians can no longer think of time as purely secular.

Time is understood by Christians in terms of God having personally participated in its flow as one of us. We view Jesus not only as "one with the Father," but also as one who walked and talked, became tired and rested. Thus, we who walk and talk, become tired and rest, can relate our experiences in life with his. We can understand our ordinary actions—the placing of our shoes on our feet, the repair of a broken chair, the throwing of a frisbee, the planting of tomatoes, the driving of an automobile—in terms of what Jesus would do were he in our place. We can also see these things in terms of what we are doing in place of him during our time on earth. Truly committed Christians delight in thinking that way about the one who claimed to be the Way. When doing so, we believe that we become one with him.

Indeed, time's meaning has undergone change, but not because time has switched its mode of behavior with the coming of the Messiah. Certainly, the physics of time has not shifted. Yet the age-old understanding of time itself underwent a radical change. Christians visualize time now in terms of the human race having awaited Christ's coming, and ourselves afterwards celebrating his presence in this world as one of us.

In our Christian awareness of time, our odyssey of meaning focuses on God's ongoing presence in this world as one of us. In a world where time itself has been graced by Christ's presence in it, our understanding of time as related to him is the primary meaning of time. All other ways of understanding time (including scientific definitions) are secondary to this central consideration.

By participating in time like we do, and by personally immersing himself in the progression of moments, one after another, God elevated time to a new realm of meaning. By becoming vulnerable to time's flow and people's whims, God through Christ has given us reasons for recognizing time in transcendent ways, endowing time with higher-than-usual definitions. One such definition of time might be: "The ongoing progression of duration where truth-expression thrives on a continuing basis."

Strange as it may seem, there has been a growing awareness among Christians that the Christ-event enables us to actually transcend time. On occasion, the present seems to coalesce with the future and the past. One senses it personally, or spots it in certain writings. In the Liturgy of the Hours, for example, we read:

Father... in Jesus Christ, Our Lord and King, we *are already seated* at your right hand.[1]

In *Silent Music* William Johnston writes:

There is a gradual revelation through the ages of a Christ who has been present in the heart of matter long before he was born as a man in Israel. It is as though the incarnation of Christ is in some embryonic way contemporaneous with the birth of the universe.[2]

Christians grow to recognize that God, in some mysterious way, dwells in us through Jesus' good works. In turn we ourselves, when prayerful, actively participate in creation with God, even in ways reaching out beyond time. Through prayer we release our grip on time to be carried away in spirit with God, so that we are in spirit "exercising care" of everything in the whole of creation through God. As Christians, we become co-creators with the Lord by way of our intrinsic identity with Christ who shares his eternal Spirit with us, taking us beyond the "befores" and "afters" of the present moment.

One rather haunting characteristic of our life experience is that, psychologically speaking, the past never truly vanishes. The past lives on in our awareness and, considered from the viewpoint of the present, influences us in shaping our future.

It could be said that the meanings behind all of our lifetime experiences will constitute the foundation for our eternity. In that state of existence we shall understand that it was God's presence—the ongoing graciousness of Life throughout our earthly adventure—that rendered creation knowable, meaningful, and filled with hope and expectation.

All our good-faith responses to the world that we knew were, indeed, responses to God.

For Christians, response to this world is response to God through Christ, who is God-become-world. For us, it can be no other way. Once God has walked this earth as a human, every person and thing in this world is related to God through Christ. Thus, as Christians who reflect on the ultimates in this life, we recognize that every event everywhere in the realm of time draws upon Christ for its fullest meaning. Nothing of this world can henceforth be correctly seen as unrelated to God. It is through this channel of awareness that people who engage in creation spirituality recognize the Truth of Christ in everything that has existence.

Relative to past, present, and future, it is easy to be impressed by events that were foretold long before their actual occurrence. Why does the human spirit regard foretold events to be of higher significance than those not foretold?

Aware of our inability to predict with accuracy most out-of-the-ordinary events, we instinctively regard accurately predicted events as somehow arriving "from beyond." This fact is of high importance in the faith-stance of maturing Christians, for it is well known that much about Jesus' life—for example, Bethlehem as his coming birthplace (Micah 5:1)—was hidden in prediction passages that can be found in the Old Testament. New Testament writers were aware of this and frequently mention Old Testament passages in their gospel accounts.

Christians eventually grow to realize that Jesus' struggle was one in which all humans participate. It is sometimes said that we, individually and collectively, bear a responsibility for the crucifixion of Jesus, even though his death occurred long ago and far away from where we presently dwell. How can that be?

The answer hinges on the understanding that Jesus, as the Christ, is eternal. Having identified himself as Truth Personified, his Truth—loving and glorious in its fullness—transcends time and place.

Sin has always been the result of people crucifying the Truth that is alive in their spirit. When sinning, we knowingly deceive ourselves, squelching within us what we know to be life-enhancing, namely the eternal Truth that is Christ.

The demeaning outcome of choices founded on falsities is that we afterwards know ourselves as being persons who turned our back on Truth. This psychologically crippling condition cries out for healing in the framework of reconciliation with Truth. In the Christian psyche, it calls for bringing into the light the truth of one's own darkness. It demands a specific naming of our wrongdoings so as to include even them in the truth of what it is that God has forgiven.

Realizations such as these enable Christians to recognize their own responsibility for the crucifixion of Jesus. In the world of meanings, where the genuine is expressive of the real, sin everywhere in every era reaches out beyond specific times and places, for it is a turning away from eternal Truth transcending time and place.

In our good-faith responses to the world, we say in effect, "Let Christ live forever, particularly in the here and now of this moment!" Thus, Christians have reason to speak of all things, places, and moments of time in terms of the Truth that the world reveals, for the Truth of Christ pervades the whole of the cosmos. It is an attribute of creation that is synonymous with "the Christness of the world," a term which, sad to say, is almost never heard.

Notes

1. Sunday, Evening Prayer II, Week IV.
2. Johnston, William, S.J., *Silent Music*, Harper and Row (1974).

For Discussion

1. Recall your day-to-day use of time across the past month. What part of this time do you feel was spent in truly proper ways?

2. Do you think that your use of time (including your sleeping) relates in any way with Christ?

3. With regard to the passing of time, which of the following words describe your feelings?
     (a) delight
     (b) anxiety
     (c) gratitude
     (d) sadness
     (e) nostalgia
     (f) anticipation

CHAPTER 33

The World as Holy Sacrament

"God's presence in this world is not like that of a foreign dignitary visiting
the squalor of a third world slum, inoculated and vaccinated against infection
and disease, well fed and with an airline ticket in his pocket ready to speed
away into the sky after saying comforting things and giving a few hand-outs.
The God who 'visits' this world stays to the death."
—Adam Ford

In chapter 12 on the subject of meaning, we considered a Papuan tribesman examining what we call a pencil, and thinking of it as a new kind of blowgun dart. We considered how people assign meanings to things in very subjective ways, and we reflected on how our world in its vulnerability stands before us, patiently awaiting our bestowals of meaning on its countless components.

With some imagination, we can proceed beyond the simple recognition of creatures in terms of their utilizations; we can relate to things in endless other ways. For example, we can pick up a pencil and mindlessly write our name. Or, again, we can pick up the pencil and reflect on how long a time humans have had to wait for its arrival. We can visualize the lumberjacks who cut the tree for the wood of the pencil, the hardworking miners who obtained the graphite, the highly trained polymer chemists who developed the synthetic rubber for its eraser, the team of men and women who manufactured its paint, the shipping department personnel, the truck drivers, and the merchant who finally placed the pencil in the hands of the customer. We can even place the wood of the pencil under a scanning microscope and discover with amazement the tubular structure of its fibers.

And, further, we can appreciate people, things, and situations associated with the world of pencils by celebrating the fact that we have fingers with which to grasp the pencil, that we have a name to write, and that we have readers willing to receive our writings, thereby rescuing us from non-recognition in the world of writing. To think in this way is to move beyond the shackles of false materialism and relate to God's mystery of creation through ordinary things and people. Picturing our world poetically with God in mind allows us to become more fully alive and *properly materialistic*, acclaiming the wonder of God by way of creation appreciated.

But are there still higher ways of visualizing the wonder of things material, of seeing God's immediate presence throughout the physical universe without engaging in pantheism? Indeed, there are such ways.

In pondering the transcendent meaning of the world's substance, Christians inevitably arrive at what we call the "eucharistic presence"[1] of Christ, a central point in their religious stance. The notion of *eucharist*, as presented in John's gospel, should catch the attention of psychologists, physical scientists, and others who navigate the world of meanings. It raises the question of just how far we can validly go toward understanding matter in a radically different way. Stated more pointedly, how can anyone with intelligence say, as numerous Christians have traditionally maintained, that bread *becomes* the living body of Christ? And that wine *becomes* the living blood of Jesus who died in Jerusalem some twenty centuries ago? Might that belief be merely symbolic, a comforting notion that people have used across the centuries for lifting their spirits?

Let us approach this subject from another direction by first considering a passage from John's gospel:

> [Jesus said]: "Whoever eats of this bread will live for ever; and the bread that I will give for the life of the world is my flesh."
>
> The Jews then disputed among themselves, saying, "How can this man give us his flesh to eat?" So Jesus said to them, "Very truly, I tell you, unless you eat the flesh of the Son of Man and drink his blood, you have no life in you. Those who eat my flesh and drink my blood have eternal life, and I will raise them up on the last day; for my flesh is true food and my blood is true drink. Those who eat my flesh and drink my blood abide in me, and I in them." (John 6:51–56)

These words attributed to Jesus are powerful, radical, and explicit. They leave little room for saying that Jesus meant the bread and wine to be merely symbolic of him. The scriptural writer was conveying that Jesus meant what he said, and that he was correcting those who misunderstood his meaning.

Several important questions arise here: Can we look at what appear to be bread and wine and understand them as infinitely more than "mere bread" and "mere wine"? Can we look upon them as becoming the body and blood of Christ at a specific moment in time? Can we do this more than two thousand years after his death?

Before continuing with this reflection, let us recall how we commonly indulge in certain experiences of multiple meaning. Indeed, we often become attached to particular things by way of the diverse meanings they hold for us. For example, an envelope might contain a folded sheet of paper—a letter. In keeping a letter received from his faraway sweetheart, a soldier feels something of her living sentiments contained in the envelope he carries. Her letter has meaning for him, a meaning that goes beyond what is signified by the paper and dried-up ink that comprise her writings.

Relative to the Eucharist, scientific tests, were they to be conducted, would reveal no important changes as having occurred in the bread and wine at the moment of their consecration. However, such a finding would prove very little. And, furthermore, it would be unscientific to insist that there can be no realities beyond that which is sensually or instrumentally perceptible.

It seems safe to say that everyone who has made the Christian leap of faith will agree that it is very plausible for one to shift one's perspectives so as to visualize specific things from different points of view. Scientists often do this. For example, they repeatedly view light in dualistic ways, as particles and as waves, with both views being experimentally verifiable.

With regard to the Eucharist, we enjoy a kind of dualistic understanding of "world," beginning with our ordinary ways of viewing bread and wine, and culminating with our understanding of God through Christ dwelling in the Eucharist. Roman Catholic Christians, for example, have traditionally viewed the substance of bread and wine as being changed into the body and blood of Christ. This occurs at a particular time at the hands of a presider, during a special moment at the commemorative meal called "Holy Mass," in remembrance of the Lord Jesus at the Last Supper. These eucharistic celebrations are always conducted in the context of a biblical story. The presider recounts of Christ: "While they were at supper he took bread, said the blessing, broke the bread and gave it to his disciples, saying: Take this, all of you, and eat it: This is my body which will be given up for you." The story itself is perceived as transcending time, as being not only told but also enacted, *as becoming alive in our midst during its telling.*

Roman Catholics, in particular, do not see the Eucharist as "representing" Christ. Nor do we believe that it signifies, symbolizes, or "means" Christ. Rather, we believe that the Eucharist *is* Christ—living body, blood, soul, and divinity. A surrender on our part is at work in this gigantic leap of faith. It is a mystery we embrace. We speak about it, listen to commentaries on it, read and write about it. Those who do not embrace the mystery as we do sometimes wonder how it can be accepted by intelligent people without violating their dedication to good reasoning. We who believe in the real presence of Christ under the appearances of bread and wine do so simply on the word of Christ. The mystery is seen as resting totally on the words he uttered. And because of who he is, we believe him unconditionally, and we go on believing because we derive our greatest happiness on earth by way of our faith in that mystery. In the minds of those who fully believe in the eucharistic Real Presence, maintaining a lesser view would be wholly unacceptable.

The concept of God's presence is of extreme importance in the world of religion. Historically, the term *real presence* has been of overwhelming concern to Christians whose spirits, like those of the ancient Israelites, have been sustained by an awareness of God's immediacy in their lives. In the Christian psyche, the Eucharist is a powerful sign of this immediacy.[2]

All in all, sacraments are like beacons to many (and perhaps most) Christians whose faith leans heavily on the weight of tradition. We understand sacraments as unique encounters with Christ, as rituals of immediate salvation centered in Jesus. This is not to say, however, that wherever sacraments are *not*, Christ's presence is lacking. For, as Karl Rahner explained, the sacrament of the Eucharist is best understood as a broadening experience, not as a narrowing one. The Eucharist sensitizes us to Christ's ongoing presence throughout the whole of creation, said Rahner. He regarded the Eucharist as *a particular instance* of the endlessly larger presence of Christ that spans the universe, which can be viewed as his *extended body*.

Teilhard de Chardin regarded the Eucharist as an instance within a single Great Communion—one in which all things everywhere are expressions of the cosmic Christ's real presence in the world. He wrote:

> ... in the Christogenesis which every Christian venerates... there appears to the dazzled eyes of the believer the eucharistic mystery itself, extended infinitely into a veritable universal transubstantiation, in which the words of the Consecration are applied not only to the sacrificial bread and wine but, mark you, to the whole mass of joys and sufferings produced by the Convergence of the World as it progresses. And it is then, too, that there follow in consequence the possibilities of a universal Communion.[3]

And he went on to write in his highly poetic mode:

> In the new humanity which is begotten today the Word pro-
> longs the unending act of its own birth; and by virtue of his im-
> mersion in the world's womb the great waters of the kingdom of
> matter have, without even a ripple, been imbued with life. No
> visible tremor marks this inexpressible transformation; and yet,
> mysteriously and in very truth, at the touch of the supersubstan-
> tial Word the immense host which is the universe is made flesh.
> *Through your own incarnation, my God, all matter is henceforth
> incarnate.*[4]

This in no way implies that recognition of Jesus in the substance
and/or form of the world itself would supplant the sacrament of the Eucharist for, after all, Jesus specifically commanded his followers to partake of the Eucharist under the appearances of bread and wine. It seems,
rather, that Teilhard cited the Eucharist as the central consideration in
carrying out the command of Christ, this leading outward into further
understanding of a God fully and immediately present throughout the
whole of creation.

Many Christians may not realize that the eucharistic presence of
Christ is not only a gift to us, but also a necessity. It is a gift, indeed, and
one of infinite worth and meaning. But a certain psychological necessity
exists here. Once God appears on earth as a human, walking and talking
with people, and once those people understand something of who Christ
is, then his total disappearance from their sight would be psychologically
devastating unless they were able to discern that, even after his disappearance from their sight, he is still with them in person through other
modes of closeness: as food in the Eucharist; as presence in Christian
gatherings; as teacher and revelation in the scriptures—all through the
indwelling of the Holy Spirit.

By virtue of who Christ is, then, his followers need to personally
sense him as fully lingering in this world. They know that he said, "If you
do not eat the flesh of the Son of Man and drink his blood, you have no
life in you." Could he have uttered those words so long ago without having made it possible for us to fulfill them today?"

If, as many Christians believe, Christ is God united to matter by way
of his living flesh and blood being manifestly in the world today, then finally all our knowledge of the world converges on him for its deepest
meaning. In the words of Teilhard, "Christ reveals himself in each reality
around us and shines like an ultimate determinant..."

Spiritual writers generally agree that Christ is immensified in our
consciousness in proportion to our Christ-like responses to this world.
In the fully Christian psyche, then, Christ and "world" go together as a

unified consideration, and "the Christness of the world" becomes a central issue in our journey through life. Down through the ages we have had poetic reflections on this truth, such as the following one, known as "Breastplate of St. Patrick":

> Christ be with me, Christ within me,
> Christ behind me, Christ before me,
> Christ beside me, Christ to win me,
> Christ to comfort and restore me,
> Christ beneath me, Christ above me,
> Christ in quiet, Christ in danger,
> Christ in the hearts of all [who] love me,
> Christ in mouth of friend and stranger.

Viewed from the perspective of Christ's extended body, the world assumes a new dignity and glory. The ships, the waves, the wind, the trees, the animals, the electric light bulbs (including even the burnt-out ones), the piano keyboards, the screws on our cameras, the wings on our bees, our automobile turn signals, the shingles on the roofs, our fingernail clippers, and even the rust on long-discarded tin cans—all of these glow with new meaning in Christian eyes. In the truly Christian awareness, each thing in its own way bears a closeness to Christ. And Christ communicates to us through each of them a mode of his abiding presence, his truth and love, his companionship and availability, his wonder, his hope, his promise. Each creature is understandable in the light of its particular charisma, in terms of what makes things fully real. It is here that we find the mystical presence of Christ, the God-among-us who is discernible in creatures everywhere being what they are—namely, outer expressions of the inner life of God.

In the maturing Christian consciousness, then, things once visualized as special for their specific substance, form, appearance, mode of behavior, or utilization possibilities now assume new meaning. We see them additionally as unique for newly discovered reasons, such as for the ways in which they express Christ the Way in terms of their own specific ways. It is precisely this quality of the world—its "Christness"—that increasingly shapes our sense of wonder as we advance in Christian awareness.

Our standards of value have shifted. Christians in time come to sense that our atmosphere is good not only because it supports biological life, but also because Christ, who is Life in its Fullness, has sanctified it by once having breathed it.[5] Motion is wonderful not only because it signifies energy, the ability to do work, but also because Christ utilized motion in walking from place to place while doing the will of his Father on earth. Proper amounts of rest and sleep are desirable not only for reasons of health, but also because Christ rested and slept. Food and water are of transcendent delight not only because they nourish our bodies, but also because the Lord Jesus partook of them while dining with beggars and sinners. He is the One who, having entered into the world, necessarily lingers, and in doing so, redefines everything everywhere. As St. Paul wrote:

> He is ... the first-born of all creatures. ... He is before all else that is. In him everything continues in being. (Colossians 1:15–17)

Ultimately, in the fully Christian outlook, Christ's living presence is at issue in every consideration of this world. And to whatever extent creatures (as God-expressions) bring Christ to mind, to that extent they serve as a Eucharist of a kind for us in a world that is sanctified by God's intrinsic presence in it through Christ.

Notes

1. Eucharist is a word derived from the Greek *eucharistos*, meaning grateful or thankful.

2. Not to be overlooked here is the fact that Christians also believe that the real presence of Christ dwells in the words of scripture, as well as in religious gatherings of Christians.

3. Pierre Teilhard de Chardin, *The Heart of Matter*, New York: Harcourt Brace Jovanovich (1976), 94.

4. Ibid., 123.

5. As explained on page 30, in every breath we take, we inhale some of the atoms exhaled by every other person who has ever lived throughout the ages.

For Discussion

1. Discuss this statement: "Once we accept Christ as alive and present in this world, then everything and every moment becomes sacred."

2. Comment on the following: "If Christ is God, and if nothing in this world is aside from God, then Christ's living presence is in some way at issue in every consideration of this world."

3. Certain persons accuse Christians who partake of the Eucharist of engaging in idolatry. How would you respond?

CHAPTER 34

Is Human Life Forever?

In today's world the scope of personal experience has certainly expanded beyond the wildest dreams of those living a century ago. Far-away experiences of others can instantly become our own with no great use of imagination. Movies and television bring us cheetahs chasing antelopes on the Serengeti plains. We experience helicopter flights over deep gorges in Utah. We swim through caverns within icebergs where, through the eyes of underwater cameras, we marvel at tile-like floors hewn by salt-water crystallization. Through underwater TV cameras operated by others, we watch penguins accelerating upward to the water's surface, emerging with sufficient speed to land feet-first on ice floes. We witness the motion of hummingbird wings slowed down to the speed of seagull wings. We see instant replays of sporting events and sometimes use them to settle questionable referee calls. Countless adventures such as these place us far ahead of previous generations in sharing our world experiences.

But toward what do these shared experiences converge? What, indeed, is humanity as a whole destined for? Addressing this subject, Teilhard de Chardin explained how humankind is being forced into closer proximity by the size limitations of the planet and by remarkable expansions in communications such that a "common soul" will increasingly come to life. He also wrote as follows:

The last blank spaces have vanished from the map of mankind. There is contact everywhere, and how close it has become.

We are moving both freely and ineluctably in the direction of concentration by way of planetization.

...The curve of consciousness, pursuing its course of growing complexity, will break through the material framework of time and space to escape somewhere toward an ultra-centre of unification and wholeness, where there will finally be assembled, and in detail, everything that is irreplaceable and incommunicable in this world. And it is here, an inevitable intrusion in terms of biology, and in its proper place in terms of science, that we come to the problem of God.[1]

Teilhard wrote of how the Divine "presses in upon us," awaiting us in created things and situations, sometimes advancing to meet us.[2] Some of the ways in which God advances to meet us are through our personal recognition of good fortune, through scriptural passages, through loneliness after sinning, through the acclaim we receive from others, through the misfortunes that make us reflect on what is important in life, and through the relentless passing of years that brings to us a sense of increasing urgency about life and death.

In reflecting on the passing of time, we eventually think of our future days on earth as shrinking in number. This becomes a driving psychological force that impels us to consider what we must let go of. The unmanageable flowing of time leads us to ask, "What is life?" in a manner quite different from the way in which biological scientists pose the question.

On reflecting deeply about our world experience, we come to realize that, in a sense, we ourselves contain the world, because the world to us is simply the summation of our expanding inner experiences corresponding to it. In the words of Carl Jung:

> The world to us is the images of the world presented to our inner senses, the received "input" that is assimilated and interpreted within us as inner experience corresponding to the reality "out there."[3]

When our earthly life expires, our sensual experience of the cosmos comes to an end. No longer will we in the same way see the sun, the moon, the forests and one another. Nor will we hear the birds, or feel the graininess of sand, or experience raindrops falling on our arms or snowflakes striking our face as we did while on earth. Thoughts like these tend to fill us with sadness. Yet there is much more to be said and there are many considerations still to be pondered. For instance, why were we so in love with this world in the first place?

Our experience of earthly life is so transcendent that we cannot help falling deeply in love with it despite all of its adversities. People of faith can say that the reason we fall in love with things and people is that they reveal God's loveliness on deeply subconscious levels. In time we come to understand that it is the vision of God we experience in them that makes it difficult to let go of them. We see something of God in their availability, their presence, their devotion, their adaptabilities, their uniqueness, their similarities and differences, and the options that they hold open to us. We recognize a loveliness suggestive of divinity in their faithfulness, their steadfastness, their suitabilities, their alluring shepherd-like qualities of leading us on in hope-expectation, their lamb-like submissiveness, their challenges that move us forward toward self-improvement, their dependability on which we rely, their companionship and their "just being there." We sometimes even delight in the splendor of their absence from places where they do not belong. Subliminally, we go about extracting from things and people something of the divine that we find to be of wonder. They linger in our memory, and the prospect of death separating us from them affects us deeply. Deep within our hearts we ask: "Lord why do you do this to us?"

Indeed, we are very concerned with our great limiting factor, our earthly mortality. After all, as people who discern the wisdom of acting today with tomorrow in mind, we naturally want to extend our present understanding of life beyond the finiteness of what we experience on earth.

Is it possible that at death we will totally cease to exist? Will there be no tomorrow-experience to life on earth? Or, as a friend of mine once joked, "Do you think that when we die we might disappointingly discover that life does not go on?"

Keeping in mind that human life is a developmental expression of nature as a whole, we can admit that what we experience within ourselves is not removed from the ways of the cosmos understood in its totality. So the question quickly becomes: Considering that the cosmos has nurtured humanity along the path of evolution from its beginnings, nourishing within our hearts a deep-seated desire never to come to an absolute end, would this entity play on us the supremely dirty trick of leading us into annihilation, with nothing of ourselves existing even to sense an answer to the question of whether Jesus' promise of life everlasting was genuine?

It would seem that our mental health, an outcome of nature's evolutionary processes at work within us, demands belief in a hereafter. Most certainly, people would be unable to function normally and adaptively in this life if they recognized themselves as destined for total oblivion. The very concept of total oblivion, in fact, seems preposterous and impossible to visualize—leaving people bewildered as they grope toward figuring out what a total-oblivion non-experience would be like.

Human life today would have questionable meaning if we understood ourselves to be alive in a steadfast movement toward nothingness. It seems certain that a psychological affliction would overtake us—one marked by hopelessness stemming from an inability to look forward to wonderful closures. Human anxiety would take many forms, ranging from deep despair to violent elimination of what is found displeasing. Morality would seem pointless, the practice of love would have low meaning, and the concept of wisdom would be vastly different from what it is today.

Personally having listened closely to the ways of nature for a very long time, and having communicated them in classrooms across four decades, I cannot imagine nature as anything other than nurturing, fulfilling, and life-enhancing in the long run. In spite of its various calamities including bodily death, nature's ongoing role as provider of opportunity remains most impressive.

People of religious faith easily recognize a caring deity at work through nature. We ask ourselves, "Would nature, or God by way of nature, continually foster a fervent hope within us for life unending, only to dismiss us into oblivion at the time of physical death?" Indeed, to people who presently live close to nature and also believe in God, the notion of a hereafter is simply a proper extension of their present faith in a nature that is rested in God.

Of course, to counter this one could argue that nature is habitually playing dirty tricks on us: allowing flowers to bloom, for example, only to afterwards fade and die. But in speaking like this, we are not telling the whole story. In that particular manner of thinking, we are not including the world of the spirit where essences remain after appearances disappear. For if we ourselves, as segments of the cosmos, can remember with fondness a Christmas toy discarded long ago, then something of its essence remains alive in the cosmos to this day. Something of it dwells in the inner workings of the world, at least by way of our spirits as we recall it today and perhaps are motivated to construct another one just like it beginning tomorrow morning.

To say, then, that death brings about our total end would be a radical step. It would mean that nature, which nurtures hope/expectation is our adversary, and that Jesus, who spoke with the highest reassurance of the hereafter, has deceived us. It would imply that the cosmos has evolved creatures like ourselves who harbor aspirations that lie beyond all possibility of fulfillment. In the final analysis, it would suggest that meanings in life are without meaning, that higher yearnings can have no fulfillment, that humans are in effect being manipulated, and that all of life is like a ruthless April fool prank.

Still, one might protest that adhering to the notion of a life hereafter is like allowing the tail to wag the dog, permitting our wishful thinking in favor of life everlasting to dictate how nature must behave. But I would caution that a dog-and-tail paradigm might not be appropriate, for such a view posits nature as fragmented, whereas nature, in truth, is quite harmonious, although not always to our liking.

Although we can sometimes fail in deciphering nature, it seems that nature, which we clearly understand as an immediate milieu of truth expression, cannot deceive us on so fundamental an issue. Can we reasonably maintain that the hope for life everlasting—a hope we find implanted deep within ourselves—exemplifies nature gone awry? If humans advance along the evolutionary track in the natural order of things, then must not nature be nurturing in some way those highest aspirations that we find within ourselves? Must not nature be performing through us an orchestration of wondrous proportions and offering a closure that extends beyond our present life?

There are many ways of thinking about this. For example, one might assert that humans represent an anomaly of a sort, an aberration or digression in nature, or that nature "outdistances" itself in the case of humans. But in speaking this way, one would be implying that nature harbors something of the metaphysical within itself, perhaps by having strayed beyond its ordinary behavior and evolving humans as if by surprise. It would be suggesting that nature, by some fluke, has behaved in other-than-normal ways when it comes to humans. However, that approach, while seeming to be novel and somewhat unorthodox, carries us rather close to what many religions have long been preaching—a dimension in nature that leads us to the doorsteps of the supernatural.

Christians understand nature's calls and their own yearnings for fulfillment as somehow blended, as harmonized in the context of the promises of Christ. We who believe in evolution view nature as inclined from its beginnings toward the emergence of Christ who would interpose himself between our highest personal yearnings and their victorious fulfillment in what he called the "kingdom" of his Father.

Understanding nature, then, *as inclusive of Christ*, Christians can visualize the universe as harboring within its makeup the qualities of sin-forgiveness and life everlasting. For if, by way of evolution, the workings of the world have brought us a Christ, then the world inclusive of him cannot be dismissed as deceptive. This is the fundamental notion at work when Christians say that Christ has "redeemed creation itself." They imply that everything in creation now rests on him for meaning.

At every turn, at every moment, we continually address the cosmic reality with hope and expectation—when searching for a restaurant or reaching for the knob that enables us to open a door. We call nature Mother and speak of Father Time, believing that the universe ultimately has our best long-term interests at heart. But does the term "our best interest" include even the phenomenon of death? Indeed, could it be that physical death itself happens to be in our best interest? Jesus once said:

> Unless a grain of wheat falls into the earth and dies, it remains just a single grain; but if it dies, it bears much fruit. (John 12:24)

Whether we view him as God-become-human or as human-become-God, Jesus Christ was undoubtedly the greatest discerner of life the world has ever known. Repeatedly, he emphasized the reality of the here-after—a kingdom in which he would live in glory with his Father and with his friends in the Spirit. Shortly before his death, he spoke of going to prepare a place for his followers. So fully has he identified with us that in dying we become like him. He spoke of himself as *the* resurrection and *the* life. And he stated emphatically: "Those who believe in me, even though they die, will live" (John 11:25).

Christians deeply believe that Christ is who he said he is, and that he is fully with us throughout our death-diminishments. We believe that nature, understood holistically as orchestrated by God, is fully our friend, and that time's very meaning is centered in Christ who, in becoming one of us, personally took part in the workings of nature.

We need to remember that, as outcome of this majestic orchestration, Heaven itself must include the fruits of our imagination, our capacity to dream really big. Our imaginations too are part of the cosmos, because they are associated with our bodies that are outgrowths of the cosmos. Heaven must be of such dimension as to satisfy our highest, grandest, and most beautiful imaginings of the truly majestic.

Death is our defining moment, our time of losing sensory touch with the world. It is our time for putting aside time, for abandoning our plans, for letting go of all our models, myths, and paradigms in exchange for the Reality in all of its Magnificence that has masqueraded behind the wondrous in all that we have ever known. In the psyche of Christ-centered people, our death-diminishments lead us ever deeper into the Christ-dimension that is already within us as we reconcile ourselves to leaving all things of this world while holding on to their highest meanings in terms of the good, the true, the beautiful. Left, finally, with ourselves at our simplest and truest—devoid of pretense, or scheme, or masquerade—we shall not shrug off, laugh away, or dismiss any falsities of ours except through God's forgiveness that will be our only recourse. As Jesus clearly taught, we shall enter God's kingdom only by way of the sheepgate, the narrow door where the Shepherd is stationed.

In accord with Jesus' message, our ultimate fulfillment will not be a direct outcome of our social status, our distinctions, our good looks, our stylish clothes, the horsepower of the automobiles we drove, the awards received, or the property we accumulated while on earth. Our knowledge and high IQs, our education and recognition for excellence will indeed count, but *only to the extent* that they have been used in the service of others, for Jesus has said that the greatest people are the ones who serve. Only those who, like himself, exercised compassion for the needy shall be lifted.

Jesus has emphatically identified with the needy of this world. In serving the needy from all walks of life, we are serving him. And, through him, we are serving his Father. As Jesus reminded us:

> Truly I tell you, just as you did it to one of the least of these who are members of my family, you did it to me. (Matthew 25:40)

Is human life forever? Jesus spoke about it many times and was most emphatic about his Father's kingdom lasting forever. We have his word that it is, indeed, eternal. We hold him to his word, and we stake our everything on that.

Notes

1. Pierre Teilhard de Chardin, S.J., *The Future of Mankind*, Harper and Row (1959). Quotations are from pages 175, 183 and 180, respectively.

2. Pierre Teilhard de Chardin, S.J., *The Divine Milieu*, Harper and Row (1960), 47–50.

3. Carl Jung, M.D., *Psychology and Religion*, Cambridge: Yale University Press (1938, 9th printing in 1955), 11.

For Discussion

1. Are you able to imagine an after-death experience of nothingness?

2. Do you believe that every moment in this life is sacred?

3. If humans advance along the evolutionary track in the natural order of things, then must not nature be nurturing in some way those highest aspirations (such as life everlasting) that we find within ourselves?

CHAPTER 35

Convergence toward Omega

"See, I am coming soon . . . I am the Alpha and the Omega,
the first and the last, the beginning and the end."
—*Revelation 22:12–13*

In this, the final chapter, I wish first to express my gratitude to my readers who have thus far read, or perhaps struggled, through this, my
odyssey of meaning. And, second, I want to suggest further readings relating to the teachings of Pierre Teilhard de Chardin and Karl Rahner,
both of whom have enriched our world with some truly deep insights.
The notes to this chapter, as well as those on page 226, cite some of their
best works.

As people who reflect on matter, mystery, and meaning, how ought
we to respond to the realization that our home planet itself is sooner or
later to die, even if only in the very remote future? As creatures who are
called to delight, how can we enjoy life on earth while remaining fully
aware that this world is ultimately to be lost? To an imaginative person,
is this not analogous to our dancing on the Titanic? Indeed, it is, and
Jesus recognized that fact when insisting that we be always ready because
we know not the day or the hour.

In spite of perennial highly charged warnings that the apocalypse is
upon us, the truth of the matter is that we do not know how or when the
earth and the human race itself will end. Some pretend to know, going
even so far as to visit with television cameras a place where, according to
their views, the actual site of a final battle between the forces of good
and evil will occur. And so, we ask ourselves, what do scripture and science say? What have biblical writers and respected spiritual scholars
such as Pierre Teilhard de Chardin, Karl Rahner, and others written?

As scientist and priest, Teilhard de Chardin focused his eschatology
on transcendent views of the cosmos from the standpoints of both science and religion. His eschatology begins with considerations of the

evolutionary tendencies in the inanimate world, called "the lithos-phere." It proceeds through the world of biological life called "the bios-phere" and comes then to the "noosphere," the world of reflective thinking or consciousness. In his view, the "hominization" of the planet that has occurred now includes imaginative reflection on the part of hu-mans, along with the fruits and products of their imaginations such as inventions, democracies, and plans for the future.

According to Teilhard, we grow not only biologically outward from the inside, but also spiritually inward from the outside by way of "world" assimilated and experience shared. This development is as yet incom-plete because self-centered humanity is capable of straying into evil and botching creation through war, terrorism, and poisoning of the environ-ment. Deeper entry into transcendent dimensions of love leads us into what Teilhard called the "Christification" of the world, which corre-sponds to the "worldification" of God through Christ. World and God become fully reconciled in the "Christosphere," the realm of love leading toward the final "Omega point."

Teilhard's Christosphere is the realm of consciousness in which "cos-mogenesis" and "Christogenesis" merge. It is the domain into which we are brought by way of our internal dynamism in full participation with the cosmic metabolism that draws us into high transcendence. In the Christosphere we utilize matter-energy and one another's talents in the flow of time with the glory of God in mind. Love of everyone and every-thing draws us toward the "Omega point," which is the fulfillment of our highest aspirations within the framework of our limitations, including experienced sufferings. Omega is that toward which all creatures gravi-tate physically, psychologically, and spiritually in meaningful diversity, and also in meaningful patterns of unity. It is the goal of all action and all knowledge, the convergence point of all that is humanly understand-able as having significance and worth.

Teilhard insisted that, when striving toward becoming ourselves at our best, we must advance together toward the fullness of our individual personhood in confluence of thought with others, with love being the energy of our unity. Individuality is never to be subverted. In this mode, cosmic consciousness will reach its point of convergence at the so-called "end of the world." Preceding this end, he foresaw an ultimate collectiv-ity of consciousness, an overall supra-consciousness working toward a single closed system or "thinking envelope" in which individuals are val-ued with God as the center of all relationships. Unlike secular humanism that absorbs persons anonymously by failing to recognize their individu-ality as expressive of God, Teilhard's "collectivity of communion" focuses on God as subject and object of love—as Love itself aglow in the world.

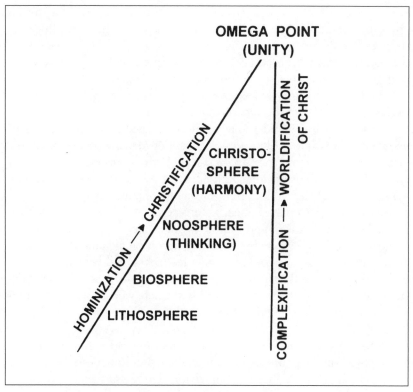

Diagram to help visualize Teilhard de Chardin's "upward swing" of the world in terms of evolution, Incarnation, and convergence.

Teilhard's Omega point is not to be understood as "a center born of the fusion of elements which it collects, as annihilating them unto itself" and dissolving away our individuality. The central focus of the Omega center is Christ who is at the heart of matter and soul, and is "the very hominization of death itself." Teilhard stressed that the Omega center is a present reality. It is currently active within our lives, giving meaning to all that we are, and to everything that we do.[1]

When writing about our present "hominized" world that is moving toward its final outcome, Teilhard seemed highly optimistic about the future, all but dismissing the possibility of nuclear war ever occurring. (He died in 1955. By the early 1980s the United States and Soviet Union possessed a total of about 25,000 nuclear bombs). It is interesting to compare his views with those of Karl Rahner, who seemed more cautious, especially regarding the ways in which "operational mankind" might entrap itself in webs of its own making. Today, when humanity is changing itself according to what many think we ought to be, lingering memories of Hitler's genetic experiments persist. These, of course, were the outcome of scientists and politicians—supposedly the super-intelligent of that era—doing what they thought ought to be done.

Rahner, in a seemingly less relaxed style, visualized a future manipulation of humanity by itself, a process that would aim, as he said, "to penetrate man's ultimate fabric in order to remodel it." He used the terms "workshop" and "factory" to express this process. He mentioned many "workshops of the factory of new human beings" that include (1) the workshop of the life sciences where hereditary factors, birth rates, IQ levels, and genetic codes are manipulated, (2) the workshops of medicine, pharmacology, and psychopharmacology that enable humans, considered as "clumsily botched together by nature," to be improved by way of artificial organs, (3) the psychological workshop where brainwashing occurs on a grand scale utilizing electric brain-stimulation to bring about feelings of well-being, (4) the sociological workshop that insists on stabilizing the earth's population at desired levels, and (5) the political workshop where world government proceeds under the control of the "super-intelligent." He wrote:

> So far this factory for the new man does not exist. But it is as though buildings are being constructed simultaneously over a great site, and one has the impression that these separate constructions will eventually grow into a single complex—into a hominized world, the one immense factory where operable man dwells in order to invent himself.[2]

In the face of these less-than-ideal prospects Rahner, however, retained a fundamental optimism, writing further:

> The Christian has no reason to enter this future as a hell on earth, nor as an earthly Kingdom of God. Jubilation and lamentation would both run counter to Christian cool-headedness. For the Christian, both he himself and his world always remain (as long as history lasts) a world of creation, of sin, of the promise of judgment and blessing, in a unity which he himself can never dissolve.[3]

Rahner wrote of death as "the zero hour through which the individual must pass on his way to the absolute future" and stated that the death of the Lord Jesus was *the solemn anticipation* of the arrival of the absolute future.[4] He maintained that Christian eschatology does not insist that things continue on after death as though we simply "change horses and then ride on." Rather, at death one's life becomes "subsumed into its final and definitive validity." Eternity should not be visualized as time unending but, rather, as *a finalized mode of spiritual freedom that was exercised in time*. Rahner insisted that eternity is understandable only through a correct understanding of spiritual freedom.

Rahner specifically referred to eternity as *the liberated final and definitive validity of our existence that grows to maturity in time*. Its growth stems from moral choices based on metaphysical insight and the radical hope of never perishing. The growth of eternity within us involves risk and loneliness, for it is based on response to goodness perceived in this world rather than on response to materials gained; on serving others rather than on being served. The growth of eternity often coincides with what others judge to be wonderful; however, it often clashes sharply with others' opinion of what is good. It sometimes calls us to deny ourselves certain immediate gains and sensual pleasures. Thus, its lack of overwhelming popularity.

Rahner insisted that the eternal is present in the absolute value of every moral decision, and that whenever one risks oneself in freedom for the sake of transcendent Goodness, one is in truth gathering time into a validity beyond the mere experience of time. As he wrote:

> Whenever a free and lonely act of decision has taken place in absolute obedience to a higher law, or in a radical affirmation of love for another person, something eternal has taken place, and man is experienced immediately as transcending the indifference of time in its mere temporal duration.[5]

Stated another way: In each "yes" we address to God, we are carried everlastingly beyond our present experiences of this world. At death we identify with and fully own the chances that we took on behalf of Goodness. And eternity is the crowning validation of the totality of these innermost "yeses."

Our Christian call is to regard our moral choices not in negative ways but, rather, positively from the viewpoint of the joy and victory that we experience through chances we take on behalf of the Wondrous. True moral decisions are impossible without love. And, thus, the adventure of attaining one's "liberated final and definitive validity" comes to completion in the *eternal crowning* of a personal love relationship between ourselves and Goodness in the highest.

There is nothing in this life to compare with the experience of being in love with Love. Happily for most Christians, we can indeed recall occasions when we have put aside our feelings and performed certain actions simply because we knew they ought to be done. Mothers who awaken at night to attend to their crying infants exemplify this type of self-abandonment. Personal services performed for the needy exemplify such generosity. The answering of the doorbell when one is occupied, the peaceful tolerating of another's idiosyncrasies exemplify Rahner's *gathering of time into a validity beyond the mere experience of time.* Always, our worldly acts of unselfish generosity bring us nearer to the Love of which the world itself is an expression.

Is it possible that the biblical "second coming" of Christ takes place subjectively in each of us at the moment of death? Might it be that, from a point of view that is subjective and yet genuine, the *eschaton* itself coincides with the end of one's own life? Thinking along these lines, Raimon Panikkar writes:

> It is significant that the most obvious meaning of the Second Coming as described in the New Testament and the Resurrection of the Flesh as maintained by the dogma of the Church, namely that it all happens in a nonhistorical context, has been almost overlooked in Christian exegesis. To put it quickly: individual Final Judgment and humanity's Last Judgment, for the individual, coalesce. The Second Coming arrives at the death of each human being.[6]

Christians easily assume that future glory in the Lord will consist in our witnessing the earth miraculously transformed into a much higher state than before—the so-called "new earth" of Revelations 21. However, another interpretation of "new earth" is that we ourselves shall be miraculously converted so as to view the same old earth in never-before-dreamed-of appreciative ways, recognizing the wonder of God and the quality of "Christness" in every person, place, or thing we have ever known or heard about. Such an interpretation of "new earth" would mean recognizing in the essence of every creature—without exception—*an outward expression of the inner life of God.*

In my view, it seems likely that a great discovery to be made in the next world will be one of high familiarity, the full realization that the Lord God was there in the wonder of every person, situation, and thing that we addressed while on earth. Heaven, it seems in some way, will be like a glorified homecoming, a transcendent awakening and reinterpretation of the highly familiar, a vivification and fulfillment of our past in the framework of reconciliation and forgiveness. Every person, every creature that we knew, every involvement in which we participated will be understood as having in its own charismatic way served as a carrier of the Shepherd's voice, as having functioned as a specific verbalization of the Word of God in a world which itself was knowable only because diverse modes of God's presence were discernible in it. And in knowing creatures, *we were in some way knowing God by way of the countless God-expressions comprising the cosmic mystique.* Surely, our openness to a proper reconciliation with every God-expression we have ever known will be an essential part of heavenly joy.

It seems that in our tremendous reinterpretation of "world," we will recognize that the reason why we are where we are is that God has all along been with us in the *ordinary things and trivial moments* of our earthly life. In our new realization, the former concept of "ordinary" will take on highest meaning.

What a revelation it shall be on the day when all eyes shall be fully opened! When athletes realize at last that it was God whom they sought when running the quarter-mile and listening to fans applaud! When farmers awaken to the fact that it was the Lord toward whom they were advancing when plowing fields and spreading fertilizer! When airline pilots and railroad conductors fully discover the Lord, the Ground of the concept of "ought," the Good whom they subconsciously sought to glorify when seeking the good of keeping their planes and trains on schedule! When mothers and homemakers understand that it was the Lord of Glory toward whom they were advancing when correcting their children, when struggling to unscrew the caps on mayonnaise jars while preparing food for others, when removing and replacing broken shoestrings and changing dust bags on their vacuum cleaners! What a glorious revelation it shall be when we discover that, poetically speaking, it was not only the hills that were alive with the sound of music, but also every other creature. We ourselves will "become as gods" in the sense of God enabling us to become as Christ.

Throughout these reflections, dear reader, I have repeatedly written of God being recognizable in every thing and person within our vast experience of "world." I have given examples of everyday encounters which, when viewed from transcendent perspectives, draw us ever forward in this world of mystery. I wish to clarify, however, that I am not implying that whenever we suffer a loss of something or someone, we thereby suffer the loss of God. Indeed, the losses we suffer in this life open new passageways to God, who is knowable also in our deprivations. By way of our identification with the sufferings of Jesus, our personal diminishments become the doors through which we discover God as present also on the shadow sides of life.

Christians view the resurrection of Christ as confirmation that their losses, accepted in good faith, are gains in the hereafter. They picture

their world, regardless of how disheveled it might become, as offering fulfillment opportunities in our milieu of hope-expression, fulfillment opportunities that hinge on the promises of the victorious Christ.

Viewed from the baseline of science and religion, Christ's resurrection implies that all of nature—every atom, photon, and possible chr-
onon throughout the totality of physical reality, including the realm of
the human brain—is playing its role in the drama of life leading toward
convergence. Without considerations of convergence, experience in this
world is merely progressive, one event coming after another, with no
crescendo, nobuildup toward a final closure, no direction toward a lasting crown of glory.

We do well, then, to visualize the whole of the cosmos through
John's vision of the New Jerusalem that is given in Book of Revelation:

> It has the glory of God and a radiance like a very rare jewel, like
> jasper, clear as crystal. (Revelation 21:11)

May you, my reader, as well as I and all of humanity reach completion in the total Omega convergence where humankind shall sparkle
like a precious jewel. May we know the fullness of the cosmic mystique
with all of creation, including the so-called "ordinary stuff" of this world,
which may not sparkle in the scientific sense of having high reflectivity
and emissivity. For that which does not physically sparkle is also precious
to God, the all-inclusive One who continually bestows on everything
that exists the sustaining crown of Presence, and the kiss of Life.

Notes

1. Pierre Teilhard de Chardin, *The Phenomenon of Man*, Harper and
Brothers (1959), 257–90.

2. Karl Rahner, S.J., *Theological Investigations*, Herder and Herder (1961),
vol. 9, 209.

3. Ibid., p. 209.

4. Ibid., p. 223.

5. Karl Rahner, S.J., *Foundations of Christian Faith*, trans. William V.
Dych, S.J., Seabury Press (1978), 439.

6. Raimon Panikkar, "The End of History," in Thomas M. King and
James F. Salmon, eds., *Teilhard and the Unity of Knowledge*, a symposium at
Georgetown University, Paulist Press (1983), n. 53, p. 127. Since retiring
from the University of California, Santa Barbara, Raimon Panikkar, an in-
ternationally known scholar, has lived in Tavertet, Spain.

For Discussion

1. Recall a "free and lonely act of decision to a higher law" that you once made. How did you feel about it at the time, and how do you feel about it today?

2. Is there a contradiction between Rahner's statement, "What a man will be for all eternity is what he has made himself" and the statement one sometimes hears from spiritual writers, "Salvation is a gift and not something we earn"?

3. Relative to Rahner's view that "operational mankind" might entrap itself in webs of its own making, what do you imagine some of these "webs" might be?

Index